Clinical Research Coordinator Handbook

Second Edition

Clinical Research Coordinator Handbook
Second Edition

Deborrah Norris

Plexus Publishing, Inc.
Medford, New Jersey

Second Printing, 2001

Clinical Research Coordinator Handbook, Second Edition

Copyright © 2000 Deborrah Norris

All rights reserved. No part of this book may be reproduced in any form or by any electronic or mechanical means, including information storage and retrieval systems, without permission in writing from the publisher. Published by Plexus Publishing, Inc., 143 Old Marlton Pike, Medford, New Jersey 08055.

Printed in Canada

Library of Congress Cataloging-in-Publication Data

Norris, Deborrah.
    Clinical research coordinator handbook / Deborrah Norris.—2nd ed.
      p. ; cm.
    ISBN 0-937548-43-X
    Clinical trials—United States—Handbooks, manuals, etc. I. Title.
    [DNLM: 1. Clinical Trials—standards—United States. 2. Clinical Trials—
    methods—United States. 3. Records—standards—United States. 4. Research
    Design—standards—United States. QV 771 N854c 2000]
    R853.C55 N67 2000
    610'.7'2—dc21
                                                             00-036695

Publisher: Thomas H. Hogan, Sr.
Editor-in-Chief: John B. Bryans
Managing Editor: Janet M. Spavlik
Copy Editor: Michelle Sutton-Kerchner
Production Manager: M. Heide Dengler
Design: Lisa M. Boccadutre
Illustrations: David Wetzel, Susan Gray, Bob Jackson,
       and David Beverage Graphic Design
Sales Manager: Pat Palatucci

Dedication

This handbook is dedicated to clinical research coordinators, worldwide, and to their commitment to the advancement of medical science while protecting patients' rights and working long hours, often without the recognition and credibility deserved.

The pivotal role they play is crucial to the development of new drugs and devices. The impact they have had, and will continue to have, on the patients participating in clinical trials makes a real difference. Clinical research coordinators are the heart of clinical research.

**In Memory of
Ray H. Weidman**

Table of Contents

Acknowledgments viii
Foreword ix
I. Introduction 1
II. Federal Regulations Governing the Obligations
    of Clinical Investigators of Regulated Articles 2
III. The Clinical Research Organization 3
IV. Duties of the Clinical Research Coordinator 6
    A. Compiling Pre-Study Documents 7
    B. Creating and Maintaining Study Files 9
V. The Creation of Study Source Documents 12
    A. Source Documents—What Are They? 12
    B. Source Document Examples 13
    C. Types of Medical Reports Generated While a Subject Is Enrolled in a Study 13
    D. Good Clinical Research Documentation Details 14
    E. Source Documentation Suggestions 15
    F. Meeting the Requirements of Source Documentation 16
VI. Obtaining Consents, Approvals, and Signatures 18
    A. Obtaining Informed Consent 18
    B. Obtaining Institutional Review Board Approvals 19
    C. Reviewing the Protocol and Obtaining Signatures 20
VII. Pertinent Forms and Study Records 21
    A. The Test Article Inventory System 21
    B. The Drug Accountability Record 21
    C. Case Report Forms 22
    D. Authorized Representative Signature Record (Site Signature Log) 24
    E. The Adverse Experience Form 24
    F. Specimen Handling and Submission Forms 25
    G. The Subject Enrollment Form 26
    H. The Site Visit Log (Monitor Log) 26
    I. The Telephone Log 26
    J. Delegation of Authority Log 26
VIII. The Pre-Study Site Visit 27
IX. Recruiting and Enrolling Subjects 28
    A. Prospective Subject Groups 28
    B. Methods of Publicizing the Study 28
    C. Presenting the Study to the Subject 30
X. Conducting the Study and Keeping Records 32
    A. Study Activities 32
    B. Subjects Lost to Follow-Up 33
    C. Withdrawing a Subject 33
    D. Site Visits During the Study 33
    E. Closing the Study and Retaining Records 34

XI.	Preparing for an FDA Audit	36
	A. FDA Inspections of Clinical Trials	36
	B. Types of FDA Inspections	37
	C. Parts of the Investigation	38
	D. Common FDA Inspection Findings	39
	E. Common Problems Identified in Audits of Clinical Trials	41
XII.	Clinical Research: Potential Liability	43
	A. Fraud	43
	1. Types of Fraud	43
	2. Who Can Commit Fraud?	43
	3. Methods Used by the Pharmaceutical Industry to Detect Fraud	43
	4. Investigator Misconduct	44
	B. Negligence	45
	C. Legal Liability and the Clinical Research Coordinator	45
XIII.	Writing the Study Summary	47
	A. Biostatistical Report	47
	B. Integrated Clinical Study Report	47
	C. Articles for Publication	47
XIV.	Achieving Credibility and Recognition as a Clinical Research Coordinator	48
XV.	Appendices	50

    Appendix I: Helpful Sources ... 50
        A. Names and Addresses of Clinical Research Resources ... 50
        B. FDA Contacts for Human Subject Protection ... 53
        C. FDA Office of Health Affairs, World Wide Web Sites of Interest for Human Subject Protection Information ... 54

    Appendix II: Forms Used During the Course of a Clinical Study ... 56
        Form 1. The Statement of Investigator Form (FDA 1572) ... 56
        Form 2. The Administrative Checklist ... 60
        Form 3. Initial Adverse Experience Report Sheet ... 62
        Form 4. The MedWatch Adverse Experience Form ... 63
        Form 5. Study Site Telephone Log ... 64
        Form 6. Study Advertisement Office Sign ... 65
        Form 7. Screen Visit Letter ... 66
        Form 8. Baseline Visit Cover Letter ... 67
        Form 9. On-Going Study Visit Letter ... 68
        Form 10. Final Study Letter ... 69
        Form 11. Generic Financial Disclosure Certification Form ... 70
        Form 12. Stop: Before Any Med Changes ... 72
        Form 13. Stop: This Patient Is a Participant in a Research Project ... 73
        Form 14. Principal Investigator Delegation of Responsibilities ... 74

    Appendix III: Individual State Regulatory Requirements for Conducting a Clinical Trial Using an Investigational Drug ... 75

    Appendix IV: Conversion Tables ... 78
        A. Conversion to Military Time ... 78
        B. Measures and Equivalents ... 79
        C. Conversion Charts for Height and Weight Measurements ... 80

    Appendix V: Glossary of Terms ... 81

Acknowledgments

It is with appreciation that I acknowledge those who have been so helpful with the revision of this handbook.

I am grateful for the patience and knowledge of many persons during the revision of this handbook. I am indebted to all of them. I want to thank everyone at Plexus Publishing, Inc. for taking the original handbook and helping to put the final pieces of it together, as well as their patience during this process, especially John Bryans, Pat Palatucci, Janet Spavlik, Heide Dengler, and Lisa Boccadutre. Thank you, Michelle Sutton-Kerchner, not only for your editorial contribution, but also for giving me the push I needed to complete the revisions.

I want to thank everyone that reviewed the handbook for their comments and suggestions, especially Kary Ehlmann, Vita Lanoce, and Kathryn Weaver. As I was doing the research for this revised edition, two people were helpful beyond measure. Andrea Marlow, the reference material for source documentation was especially instrumental in the development of that chapter, thanks. Ellen Metivier, your input on the Delegation of Responsibilities Form is greatly appreciated. Deep appreciation goes to J.D. Davidson; thank you for taking time from your busy schedule to write the foreword. Thank you for your honest comments. Nothing has been of greater importance than the communications I received from those clinical research coordinators who have purchased and used the original handbook. For technical expertise about the legal portions of this handbook, I went to my ultimate informed source, Francis X. Sexton, Jr., Esq. Any errors or oversights are mine. Thanks David Wetzel, Bob Jackson, and Susan Gray for the original illustrations, which were also used in the Japanese translated versions. They look great in this revised version, too.

Larry Liberti and Sue Dalton, the vision continues to grow … thank you for believing in this handbook from the start in 1992 and for coordinating the publishing in Japan. It's come a long way since then!

Moreover, I have relied at every point, as I have for many years, on the wisdom and judgment of my foremost sounding boards, Justin and Frank. Thank you both for your support and love.

Foreword

In this short book, you will find a straightforward explanation of clinical research and the role of the coordinator in such research. You will learn that, by following the details, paying attention to them, and maintaining an attitude of strict integrity, you can accomplish a creditable clinical research project.

Deborrah Norris has been doing clinical research longer than she will want me to admit. I have known her from her days as a college student, where she further honed her writing skills. She has always been a diligent worker who maintains details very well. A strong work ethic and innate sense of integrity have been hallmarks of her work. A candid way of getting to the root of issues is manifested in this handbook. The humor throughout furnishes a welcome addition to the presentation.

Welcome to the world of clinical research!

J.D. Davidson
Dean of Humanities, Arts, and Social Sciences
Utah Valley State College
Orem, UT

I. Introduction

The purpose of this handbook is to outline the various functions required of the clinical research coordinator (CRC), research nurse, research assistant, or medical assistant involved in the conduct of a clinical study for an investigator on behalf of a pharmaceutical company or device manufacturer and contract research organization (CRO)/site management organization (SMO) (the sponsor).

The complete and accurate accumulation of information specified in the study protocol and the subsequent accurate transfer of this information to the case report form (CRF) is vital to the successful progress of a clinical study. As coordinator of this operation, the CRC plays an integral role, often without the recognition and credibility they deserve for their dedication to this process. This handbook describes the various duties of the CRC and outlines the regulatory requirements that MUST be followed when conducting a clinical study. The documentation that is required before the conduct of a study is described, and prototypes of forms, checklists, and letters are included. In addition, the topic of legal liability as it relates directly to the CRC has also been incorporated. A glossary of terms is located at the back of this handbook to aid in the understanding of the language of clinical research.

Ultimately, it is the close communication among the investigator, CRC, and the sponsor that will ensure the timely and successful completion of the clinical study. Studies performed in accordance with their approved protocol, with the guidance of the sponsor, and under the supervision of a well-informed clinical research associate (CRA), will meet the goals of good clinical practices (GCP) as established by the Food and Drug Administration (FDA).

This handbook is written both as an introduction for the CRC who has not been involved in extensive clinical trials and as a refresher for the experienced CRC. In all cases, we are confident that the reader will gain a more detailed appreciation of the clinical study process.

II. Federal Regulations Governing the Obligations of Clinical Investigators of Regulated Articles

A clinical study is an experiment that has been designed to prove a hypothesis: for example, that Drug A is as safe and effective as Drug B; that a drug's pharmacokinetics remain unchanged in subjects with renal insufficiency; or that concomitant administration of an antacid with the drug has no effect on its absorption or other kinetic parameters. As with any other scientific endeavor, the conduct of a clinical study demands that good scientific methods be adhered to in order to ensure the highest quality results. Over the past several decades, clinical researchers have developed standard methods that, when implemented in a clinical study, contribute to its smooth progress. These methods may include the use of recog-

nized screening criteria for assessing subject eligibility, such as the use of the DSM-IIIR to categorize mental illnesses or the use of American Rheumatology Association (ARA) criteria for determining the extent of involvement in subjects with rheumatoid arthritis. Other elements that add to the strength of the study include adherence to a well-defined clinical protocol, the selection of appropriate tests to assess safety and efficacy, the use of well-designed CRFs to ensure unbiased data collection, and the use of appropriately applied statistical tests for the analysis of the data.

Collectively, these elements form the basis for the concept of GCPs. Good clinical practices have been promulgated by the FDA with input from researchers, statisticians, clinical personnel, regulatory and legal specialists, and the pharmaceutical industry. They provide guidelines for essentially all aspects of conducting a well-designed and well-controlled clinical study. While it is the sponsor's responsibility to ensure that a study conforms to the guidelines set forth, the CRC and investigator must also assist the sponsor in adhering to these guidelines.

The regulations and guidelines of GCPs are detailed, and the reader should recognize that the elements described in this manual are rooted in these concepts. The Code of Federal Regulations (CFR) 21CFR Part 50, "Protection of Human Subjects," and 21CFR312.50, Subpart D, "Responsibilities of Sponsors and Investigators," describe in detail the regulations set forth by the FDA regarding the testing of drugs in humans. Further FDA-related helpful sources can be found in Appendix I, Parts B and C.

III. The Clinical Research Organization

A large portion of clinical studies sponsored by the pharmaceutical industry are implemented and monitored through contractual agreements with specialized providers of clinical services collectively known as CROs/SMOs.

Contract Research Organizations and Site Management Organizations

What They Are and What They Do

In the past, pharmaceutical companies supervised their own clinical trials, from the discovery of the compound, to animal testing, to the preliminary testing of human subjects, to the larger clinical trials involving hundreds or thousands of patients. To assist pharmaceutical companies, CROs/SMOs have developed to handle all of that and more.

Contract research organizations are composed of companies that run parts or all of the clinical trials. Site management organizations generally are a group of physicians willing to offer their services to conduct the actual clinical trials. These organizations arose to fill a need, to assist the pharmaceutical companies in the drug development process. According to an article in *The New York Times* published in May 1999, these organizations evolved to elevate the enormous pressure that began around 1992 from managed care companies and health insurers on drug companies to hold down prices. The Federal FDA encouraged development of these companies in order to speed up its approval process. The average review time for new drug applications (NDAs) dropped from more than twenty-two months in 1990 to just over fourteen months in 1997, according to the FDA.

Pharmaceutical companies found they had to put more effort and revenue into finding drugs to develop, while at the same time cutting their development costs. For many pharmaceutical companies, their solution was to dismantle all or part of their clinical research monitoring departments that ran the clinical trials, turning the work over to smaller companies that emerged to fill the industry's needs.

Contract research organizations can assist the pharmaceutical companies by:

- ✔ designing the protocol, CRFs, and source documents required for each clinical trial;
- ✔ locating principal investigators (PIs);
- ✔ assisting in potential patient recruitment efforts;
- ✔ statistically analyzing the clinical trial data;
- ✔ meeting with the FDA;
- ✔ preparing the enormous amount of paperwork required by the FDA for submission before the test article can be approved;
- ✔ writing the scientific papers for publication and presentation at national and international scientific meetings;
- ✔ conducting periodic auditing of investigator sites to ensure the study is being conducted according to the protocol, the CODE OF FEDERAL INVESTIGATIONS, and GCP;
- ✔ selecting and coordinating central laboratories; and
- ✔ selecting and coordinating independent institutional review boards (IRBs).

Site management organizations can assist the pharmaceutical industry by:

- ✔ providing a national network of physicians experienced in conducting clinical trials;
- ✔ providing a national network of physicians in a specific area of expertise, such as neurology, cardiology, endocrinology, or urology;
- ✔ coordinating the completion of the required regulatory documents for their sites;
- ✔ negotiating the budget for their sites;
- ✔ handling the financial aspects of the clinical trial for their sites, invoicing the sponsor, and providing payment to the sites;
- ✔ providing tools for their sites to meet enrollment commitments;
- ✔ providing their sites with a set of standard operating procedures (SOPs); and
- ✔ developing study-specific source documents for their sites.

In addition to the above services provided, several CROs/SMOs have a Phase I unit, on-site or off-site, where they can conduct clinical trials introducing the test article for the first time to healthy human subjects.

The size of these organizations varies from large corporations, to groups of physicians conducting clinical trials in a specialized area such as neurology and urology, to a rolodex of the names of physicians interested in participating in clinical trials, to a directory of physicians by specialty, to a Web site listing clinical trials looking for PIs, to a company of one person working out of her/his home to develop a drug.

Due to the nature of clinical research, CROs/SMOs will continue to provide the pharmaceutical industry with the necessary support needed to conduct clinical trials. The CRC and PI play pivotal roles in ensuring the smooth conduct of a study. By understanding all elements of the clinical protocol, the CRFs, and the study design, the CRC, PI, CRO/SMO, and sponsor will be able to work together to efficiently meet the study goals.

IV. Duties of the Clinical Research Coordinator

The CRC or research nurse plays a pivotal role in the efficient progress of the clinical study. The CRC is often responsible for organizing the documentation and files pertaining to a study and for coordinating the subsequent activities of the investigators and subjects. The responsibilities of the CRC will vary at each investigational site, but may include the following:

- ✔ to review and familiarize themselves and other staff with the protocol;

- ✔ to provide prospective investigators with a copy of the protocol;

- ✔ to notify the sponsor and CRO/SMO of investigators' interest in participating in the study;

- ✔ to prepare a grant (if needed) in conjunction with the sponsor;

- ✔ to compile the required information for the sponsor and CRO/SMO;

- ✔ to review FDA requirements and guidelines;

- ✔ to prepare for the IRB meeting, including preparation of the Informed Consent document (to be forwarded to the sponsor before the IRB meeting) and other pre-study documentation, preparation of the meeting agenda, and mailing of the IRB meeting notice to investigators;

- ✔ to attend the IRB meeting and to have proposed IRB approval letters ready (both the PI and the CRC should attend, if possible);

- ✔ to identify the study site's contact person;

- ✔ to establish the need for a Letter of Indemnification from the sponsor;

- ✔ to schedule on-site visits with the sponsor;

- ✔ to participate in the investigator meeting;

- ✔ to prepare additional study records, forms, and letters as directed by the CRO/SMO or sponsor;

- ✔ to set up files;
- ✔ to enroll subjects;
- ✔ to oversee study activities, including the secure storage of the test article;
- ✔ to maintain accurate and complete records; and
- ✔ to close the study with the sponsor and to store the study records appropriately.

A. Compiling Pre-Study Documents

Before subjects can be enrolled in a study and before test articles can be shipped, several pre-study documents must be compiled by the CRC or investigator and sent for approval to the sponsor. These are:

Form FDA 1572. This is the Statement of Investigator form, which summarizes what the FDA requires for an acceptable clinical study. It must be signed by the PI. In addition, all sub-investigators (if any) who are participating in the study and who are authorized to administer the test articles must be listed. In most instances, the sponsor determines which study personnel they require on the FDA 1572. The actual form, along with a sample of a completed version, is provided in Appendix II (Form 1). Investigators who sign Form FDA 1572 make the following commitments:

- ✔ to conduct the study according to the protocol (Changes made to the protocol can be made after notification to the sponsoring pharmaceutical company, except to protect patient safety. Investigators who deviate from the protocol for any reason need also to notify their IRB in writing of the deviation.);
- ✔ to personally conduct and/or supervise the research protocol;
- ✔ to inform patients that they are participating in a research study;
- ✔ to abide by the IRB and Informed Consent regulations;
- ✔ to report all serious adverse events according to Federal Regulations to the sponsoring pharmaceutical company and the IRB;
- ✔ to read and understand the *Investigator Brochure* for the investigational material being tested;
- ✔ to ensure that other personnel involved with the study meet these commitments;
- ✔ to maintain accurate study records and make these records available for inspection by the appropriate agencies (FDA and the sponsoring pharmaceutical company and their representatives);
- ✔ to ensure initial and continuing review and approval of the research study by the IRB; and
- ✔ to comply with the Federal Regulations for investigators in 21 CFR 312.

Note: the PI may delegate some of these responsibilities to the CRC but, ultimately, the PI is responsible for ensuring that they are done correctly.

Curricula Vitae (CV). These are required of the PI and all the sub-investigators listed on Form FDA 1572. Curricula vitae should be updated every two years, and signed and dated to show that they are current. A copy of the physician's medical license as well as proof of malpractice insurance may be required by some sponsors. In addition, CVs of the CRC and the laboratory director, if appropriate (i.e., when a central laboratory is used by a sponsor), may also be required.

Lab Certification. A copy of the laboratory license, laboratory certification, and the normal laboratory values for the laboratory to be used must be on file.

Signed Protocol. This must be signed and dated by the PI.

Financial/Certification Disclosure. The following list provides the necessary financial certification/disclosure requirements.

- ✔ Every PI needs to certify/disclose if she/he has a financial interest in the sponsoring pharmaceutical company or in the investigational material being tested.
- ✔ Certification is required for each PI and all subinvestigators listed on Form FDA 1572.
- ✔ This certification encompasses not only the investigator, but also her/his spouse and each dependent child.
- ✔ The certification is applicable for the time during which the PI is conducting the study and for one year following completion of the study (i.e., after enrollment of all subjects and follow-up subjects, in accordance with the protocol). The sponsoring pharmaceutical company should be notified in writing of any change in the accuracy of the certification during the specified time frame.
- ✔ Any financial interest in the sponsoring pharmaceutical company or in the product being tested needs to be promptly disclosed to the sponsoring pharmaceutical company.

Specifically, each PI must certify/disclose that she/he:

- ✔ has not entered into any financial arrangement with the sponsoring pharmaceutical company, whereby the outcome of the clinical study could affect her/his compensation (e.g., bonus, royalty, or other financial incentive);
- ✔ does not have a proprietary interest in the study material (e.g., patent, trademark, copyright, licensing agreement, etc.);
- ✔ does not have a significant equity interest in the sponsoring pharmaceutical company, not applicable to the material being tested;
- ✔ has not received significant payments from the sponsoring pharmaceutical company with a monetary value in excess of $25,000, other than payments

for conducting the clinical study (e.g., grants, compensation in the form of equipment, retainers for ongoing consultation and honoraria paid to the investigator or to the institution in support of the PI's activities).

Other documents that are required to be on file before starting the study include the IRB approval and membership list, the budget for the study, the Letter of Agreement between the sponsor and CRO/SMO or investigator, and the approved Informed Consent Form. The sponsor will inform the CRC of the documents needed at each step of the study. The Administrative Checklist (see Appendix II, Form 2) provides a mechanism to determine if appropriate documentation has been obtained and completed.

B. Creating and Maintaining Study Files

The Study File. Once all pre-study documents have been compiled, the Study File should be created. This file will be maintained throughout the clinical study and will eventually contain samples of all forms and reports to be completed during the course of the study, in addition to all pre-study documents.

Initially, the following materials should be included in the Study File:

- ✔ Pre-study documents, including
    - ✔ Form FDA 1572,
    - ✔ current, signed and dated (within two years) Curricula Vitae of investigators, CRC, and laboratory director,
    - ✔ current medical license,
    - ✔ proof of malpractice insurance (if required),
    - ✔ Financial Disclosure Form,
    - ✔ Investigational New Drug Application (IND), safety letters, if applicable,
    - ✔ laboratory certification and normal ranges for laboratory values,
    - ✔ Protocol (including Signature Page and Synopsis),
    - ✔ IRB membership list,
    - ✔ IRB approval,
    - ✔ approved Informed Consent Form,
    - ✔ advertising;
- ✔ *Investigator Brochure*;
- ✔ Test Article Inventory and Drug Accountability Record;
- ✔ Delegation of Responsibilities Form;
- ✔ CRFs and Adverse Experience Forms;

- ✔ Authorized Representative Signature Record;
- ✔ Specimen Submission Records (if required);
- ✔ Site Visit Log;
- ✔ Telephone Log;
- ✔ IRB Fee Letter;
- ✔ Letter of Indemnification; and
- ✔ all correspondence to and from sponsor.

The CRA will review the Study File with the CRC at the beginning of the study to ensure that all of the required documentation is present.

As the study progresses, additional documents will be completed. These (or copies) should be retained in the Study File. Examples include:

- ✔ amendments to the Protocol;
- ✔ executed Informed Consent Forms;
- ✔ laboratory test results;
- ✔ shipping invoices for all test articles, Returned Goods Form;
- ✔ ongoing correspondence;
- ✔ other source documents;
- ✔ miscellaneous sponsor forms;
- ✔ revised Informed Consent Forms; and
- ✔ correspondence to and from the IRB regarding the progress of the study, Protocol deviations, and submission of IND Safety Reports.

If a copy of a form is kept in the file, the original should be readily available and its location should be noted as a memo to the investigator file.

The Subject Workfolder. Subject Workfolders must also be compiled for each subject enrolled in the study. When correctly organized, these greatly facilitate the proper collection of study data and form the core of the clinical database. Each Workfolder should contain the following, as appropriate:

- ✔ Study Design Flow Chart;
- ✔ Screening Sheet (Inclusion/Exclusion Criteria);
- ✔ Informed Consent Form (two signed copies—one for the research chart and one for the subject);

- Laboratory Requisitions Packet;
- Data Collection Form from CRF;
- specimen labels;
- chest radiographs (if required);
- disbursement requests (compensation vouchers);
- a "Stop!—Research Subject" sign (see Appendix II, Forms 12 and 13);
- clinic statements;
- synopsis of Protocol;
- letters to the subject;
- worksheet for outside departments; and
- Medical Release of Information for medical records.

Each document in the Subject Workfolder should be identified with a unique subject number. For example, 103A-207 may indicate Protocol 103A, site number 2, patient number 7.

The Audit File. At the conclusion of a study, an Audit File will be created. This will contain portions of the Study File, including:

- Site Visit Log;
- Telephone Log;
- Protocol (extra office copy, including amendments);
- Informed Consent Form;
- Drug Receipt/Return Log;
- Subject Drug Dispensation Log;
- pre-study documents;
- correspondence; and
- Financial Disclosure (See Appendix II, Form 11).

V. The Creation of Study Source Documents

Documentation of clinical research projects is undeniably one of the most important aspects of conducting a clinical study; yet, it is one of the major deficiencies found by FDA auditors. A common source of disagreements about what constitutes source documentation continues to be an ongoing dialogue between sponsors'/CROs'/SMOs' monitors, and site personnel. Source documents confirm that the data were accurately reported and that the study was conducted according to the Protocol.

The International Conference on Harmonization (ICH), the Code of Federal Regulations, and GCP require the following:

1. written, informed consent by all subjects prior to the implementation of any study-related procedures; and
2. accurate, complete, and appropriate clinical research documentation.

A. Source Documents—What Are They?

The first recording of any observations made or data generated about a study subject during her/his participation in a clinical trial is source documentation. Source documentation is the foundation of all clinical studies. Source documents confirm the completeness and accuracy of data collection and show evidence that the study was conducted not only according to the protocol, but also ethically.

CFR 312.62 (b) states: "An investigator is required to prepare and maintain adequate and accurate case histories that record all observations and other data pertinent to the investigation on each individual administered the investigational drug or employed as a control in the investigation. Case histories include the CRFs and supporting data including, for example, signed and dated consent forms and medical records, including, for example, progress notes of the physician, the individual's hospital chart(s), and the nurses' notes."

ICH GCP 1.52 states: "Source documents are the original documents, data, and records (e.g., hospital records; clinical and office charts; laboratory notes; memoranda; subjects' diaries or evaluation checklist; pharmacy dispensing records; recorded data from automated instruments; copies or transcriptions certified and verification of their accuracy; microfiches; photographic negatives; microfilm or magnetic media; x-rays; subject files; and records kept at the pharmacy, at the laboratories, and at medico-technical departments involved in the clinical trial)."

Copies of the CRFs ***are not*** source documents. *The sponsoring pharmaceutical companies' CROs/ SMOs have been providing clinical research sites with source documents that are study specific.* While this helps the CRC, generally, it is better to create source documents that are study specific and to mirror the information in the CRFs and reflect the standard operating procedures at your site.

Anyone involved in the clinical study can create source documents. This may include, but is not limited to, the PI, the sub-investigators, the CRC, the floor nurse (if an in-patient study), the physical therapist, and so forth. The pharmacist who records the dispensing of the study drug creates a source document. A research subject can create source documents when they complete study-specific diaries.

B. Source Document Examples

- A record release form with the subject's signature and date;
- letter of referral from the subject's primary physician to the PI for study screening;
- screening or intake forms;
- medical history, including demographic data and documentation of inclusion/exclusion eligibility;
- original, signed, and dated Informed Consent Form;
- amendments to the original Informed Consent Form, signed and dated;
- physical exam notes;
- progress notes;
- study-specific flow sheets;
- study-specific checklists;
- adverse event list;
- subject diaries;
- concomitant medication list;
- admission and discharge summaries;
- correspondence to and from a study subject; and
- death certificate.

C. Types of Medical Reports Generated While a Subject Is Enrolled in a Study

- Laboratory: serology; chemistry; hematology; microbiology; urinalysis;
- ECG reports;
- MRI/CT scan reports;
- radiology reports;
- pathology reports;
- admission summaries; and
- discharge summaries.

Source documents should be accurate and complete enough to permit the entire reconstruction of the date in the unlikely event of losing the CRF. Information pertinent to the subject's participation in the study should be included. Variables, such as safety data and essential effectiveness, will determine the outcome of the clinical trial and should be documented. Adverse events, treatment, and follow-up should be described well. All communication with the sponsor/CRO/SMO regarding approved Protocol waivers or deviations should be noted. (Don't forget to notify the IRB in writing of any waivers or deviations from the approved Protocol.)

It has been reported by experienced CRCs that every hour spent with a study subject generally requires an additional hour spent documenting and completing paperwork associated with the interaction. Documentation should be practical, limited to the essential, easy to routinize, and standardized but flexible enough to include unusual circumstances.

D. Good Clinical Research Documentation Details

- ✔ There should be a record of every visit and conversation with the subject.
- ✔ Electronic messages and facsimiles should be printed and filed in the site regulatory binder.
- ✔ Logs of procedures should exist.
- ✔ Records of calibration of study-required equipment, including temperature logs, should be kept and filed appropriately.
- ✔ Screening and recruitment logs (including computer-generated lists of potential study subjects) should be retained to demonstrate that inclusion/exclusion criteria were performed according to the Protocol.
- ✔ Site visit monitoring logs should be kept to document the purpose and frequency of the monitoring visit.
- ✔ Records of missing or unobtainable data required by the Protocol should be explained appropriately.
- ✔ If information must be added to a previous entry, it should be inserted as a late entry or addendum at the end of the text in the medical chart. It should not be squeezed in between previously written notes.
- ✔ If data must be changed or clarified, the industry standard is to put a single line through the original entry, change the data, and initial and date the change.
- ✔ Any changes to the data should be evident in an audit trail. Explanations added to the progress notes may be required.

A source document is ***always*** a source document. It may seem a good idea to recopy a source document, instead of crossing out yet another entry and correcting it, *but* there is always the possibility of a transcription error. In addition, these and similar attempts to obliterate or destroy original data are *not* acceptable. To an auditor, these redactions may elicit suspicion. Is the motive fraud or neatness? Once redacted, the data is no longer a source document. ***If it is not documented, it did not happen!***

E. Source Documentation Suggestions

The *Compliance Program Guidance Manual* used by the FDA to audit an investigative site is an excellent resource for source documentation. That, along with the industry standard, provides the rationale for the following documentation to support CRF data:

Notation that written, informed consent was obtained and that the consent form was dated and signed by the subject (or the subject's representative, if applicable) prior to the performance of any study-related procedure. It is generally a good idea to also note the time at which informed consent was obtained. This will confirm that study-related procedures, such as laboratory samples, were obtained after informed consent was obtained. If an investigator's signature is required on the consent form, and the investigator did not sign on the same date that the patient did, make sure the actual date that the investigator signed the informed consent is used. Note in the source the reason for the later date. ***Under no circumstances should a CRC enter patient's initials on the consent, or enter the date for the patient.*** The patient should perform these tasks by her/himself.

Record the date of entry into the study, the sponsor's Protocol number, and the subject number.

The subject's diagnosis and current physical status prior to the initiation of any study-related procedures should be noted. Include details of medical history, checking the inclusion/exclusion criteria.

Record current medications, as well as medications discontinued within the past 30 days. Review the Protocol to determine if there are restricted or excluded medications and the wash-out period required by the Protocol. It is a good idea to ask potential subjects to bring their actual pill bottles to the screening visit. This way you can be sure the medication information being provided by the subject is accurate.

Document the name and/or number of the study drug dispensed, as well as the dosing information. Note the total amount of study drug dispensed at subsequent visits; include the amount of study drug returned by the subject. If there is a discrepancy, note the reason for it.

For example:

> *Patient #1234 did not return study drug dispensed on 7/10/99, inadvertently discarded empty bottle. Patient instructed on the importance of returning study drug, even if the bottle is empty.*
>
> *Patient #4562 missed two doses of study medication while on vacation. Patient instructed on the importance of taking the study drug according to the directions.*

Patient #0001 did not return unused study medication dispensed on 5/8/99. Sponsor was contacted and gave approval to dispense additional medication. Patient will return unused study drug at the next visit. Patient instructed on the importance of returning unused medication at each visit.

Patient #6534 was screened for Protocol NA#0099 today. Informed consent was obtained prior to the initiation of any study-related procedures. The patient did not have any questions regarding the study or the study requirements. The patient understands his/her role as a study participant. The patient meets inclusion/exclusion criteria. Next visit scheduled for 2/10/00.

Document the dates and results of study-specific evaluations and procedures. Any deviations from the Protocol must be documented and reported to the sponsor and the IRB.

Record all adverse events and complaints noted by the subject during the study and for the appropriate time specified by the Protocol. Include the last dose of study drug taken by the patient and any medical intervention or action taken to treat the adverse event or complaint.

In study progress notes, record the subject's condition during and/or after treatment.

Note the final disposition of the subject and the subject's status at the time of study termination. This includes premature study termination. Make sure you note the reason for premature study termination. There is a difference between study drug termination and study termination. While a subject may be terminated from study drug treatment, the subject may not be terminated from the study. This is study- and sponsor-specific.

The most frequent cause of disagreements between a CRA and a CRC has been reported as the inadequacy of source documentation.

F. Meeting the Requirements of Source Documentation

- ✔ Study-specific labels attached to progress notes;
- ✔ Protocol-specific flow sheets;
- ✔ Protocol-specific checklists; and
- ✔ Protocol-specific templates created by the site or in some instances the sponsor to collect the data required by the CRF.

A discussion between the sponsor and/or the CRO/SMO concerning their expectations regarding source documentation prior to the initiation of the study is essential. This will avoid problems in the future. It is a good idea to obtain written requests for any requirements that deviated from your site's standard operating procedures. In advance, evaluate the source documentation requirements, and evaluate the extent to which you can comply. Do not hesitate to negotiate what is acceptable to all parties concerned. Educate the medical record department about the importance of research records.

Document any violations and/or deviations from procedures specified by the Protocol and the reason. This represents changes made to the Protocol that have not been approved by the sponsor or the IRB. Since they may affect risk to the research subject and will impact the integrity of the data, violations and/or deviations play an important part in Warning Letters issued by the FDA.

All allied health professionals involved in the conduct of clinical research are responsible for conducting research ethically. The complete, accurate, and thorough documentation of the research subject's participation in the Protocol enables the research site to demonstrate the requirements for GCP were fulfilled during the conduct of the research study.

VI. Obtaining Consents, Approvals, and Signatures

A. Obtaining Informed Consent

Law requires obtaining informed consent from each potential subject before that subject may participate in a clinical study or be screened for the study with clinical or laboratory procedures not normally carried out on similar subjects. The Code of Federal Regulations (21 CFR, Part 50, Subpart B) explains the details of informed consent.

The design and construction of an Informed Consent Form is the responsibility of the PI, but the sponsor usually provides guidance in preparing the document and, possibly, the completed sample documents. In addition, the Informed Consent Form must be reviewed and approved by the IRB at each participating institution *before* it can be used.

One of the most important responsibilities of the CRC is to make sure that every potential subject (or that a subject's duly authorized representative) reads, understands, and signs an Informed Consent Form before participating in a study. Furthermore, the CRC must be sure that all signatures, dates, and other required information have been obtained.

The Informed Consent Form is designed to give the subject or a legal representative *easy-to-understand information* regarding the following aspects of the study:

✔ general design and purpose of the study;

✔ total length of the study;

- number of visits required for the study;
- total number of subjects in the study and number at the subject's particular site;
- benefits of the study treatment;
- risks of the study treatment;
- side effects associated with the study treatment;
- specified procedures unique to the study and risks involved;
- that the subject may get a placebo and not active treatment (if applicable);
- alternative therapies that the subject may be offered;
- right of the subject to withdraw with no penalty;
- who to contact 24 hours a day in case of questions or emergencies;
- confidentiality of subject records;
- costs and compensation; and
- compensation or medical treatments available for any research-related injury.

The rationale and basic content of informed consent have been set forth by FDA regulations. These regulations are described in the CFR, which may be obtained from the FDA (see Appendix I for contact information). While informed consent has traditionally been obtained by providing each subject with a written description of the elements of the study, some organizations are using novel approaches to educate prospective subjects. One of the most intriguing methods involves the presentation of the informed consent information in a video format, in conjunction with the written document. This approach is particularly useful for demonstrating difficult-to-explain procedures (such as cardiac catheterization, a novel surgical technique, and how to complete a patient diary), and gives the subject a visual opportunity to assess the risks and benefits associated with the study.

B. Obtaining Institutional Review Board Approvals

An IRB has at least five members with a diverse background of experience. The diversity of the IRB members' backgrounds will ensure cultural, educational, and racial diversity reflecting that of the local community. Members may include physicians, nurses, clergy, social workers, lawyers, and ethics experts. This enables the board to make an assessment of the validity of the study objectives. The Code of Federal Regulations (21 CFR) Part 56 reviews the regulatory considerations associated with IRBs.

Approval of a Protocol by the IRB is required in the following three situations. In each case, proper documentation for the approval is required.

Initial IRB Approval. Generally, initial review and approval by the IRB of the Protocol, the investigator, sub-investigators, the site(s), and the study site's Informed Consent Form and advertisement (if applicable) is required before enrollment can begin.

Until the investigator receives written IRB approval, no subject can undergo any procedures for the purpose of determining eligibility for the study.

Protocol Amendment IRB Approval. Should an amendment to the Protocol become advisable, IRB approval must be received before the amendment can be implemented, in most cases. In emergency situations, the amendment may be implemented first and the IRB notified within 10 days.

Continuing IRB Approval. Continuing IRB approval must be obtained for studies that last for more than one year, or other specified period. This will consist of documentation of the IRB's annual review and continuing approval of the Protocol.

Continuing IRB approval is obtained minimally on the annual anniversary date of the original approval and yearly thereafter, or as specified by the IRB, for as long as the study continues. Your site may also be audited by the IRB to ensure the study is being conducted properly. If the IRB requires a fee, a note describing this fee (known as the "IRB Fee Letter") should be included in the Study File.

C. Reviewing the Protocol and Obtaining Signatures

Each clinical study must have a written Protocol, which is the document that gives specific instructions for the conduct of the study. Approval of the Protocol by the IRB must be obtained *before* any subjects are allowed to participate or are requested formally to consent to participate in the investigation.

In addition, the PI must submit significant changes or amendments to the Protocol to the IRB if such a change increases the risk to subjects or adversely affects the validity of the investigation or the rights of the subjects. The investigator must obtain approval by the IRB before such change or deviation is implemented. When the change is made to eliminate or reduce the risk to subjects, it may be implemented before the review or approval by the IRB. In these instances, the investigator is required to notify the IRB, in writing, of the change within ten days after implementation.

All changes or revisions to the Protocol, and the reasons for those changes, must be documented, dated, and maintained with the Protocol.

Before the site initiation visit with the sponsor, the investigators and CRCs must read the Protocol, should highlight important topics, and identify any questions that need clarification by the sponsor or CRO/SMO.

VII. Pertinent Forms and Study Records

A. The Test Article Inventory System

The Test Article Inventory System will dictate where test articles are to be stored and who will keep track of the amount of test articles received, dispensed, and returned to the sponsor. The inventory system must ensure that:

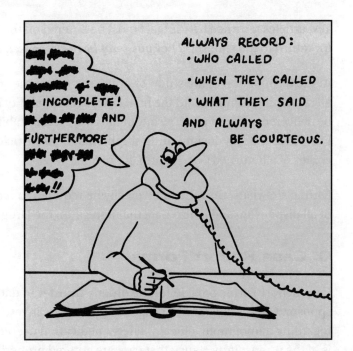

- ✔ adequate supplies are always on hand;
- ✔ test articles are stored securely and in the proper environment (e.g., the refrigerator);
- ✔ the Drug Accountability Record is properly maintained;
- ✔ there is agreement between the number of test articles received versus the number of test articles dispensed (as recorded on the Drug Accountability Record and CRFs) versus the number of test articles on hand;
- ✔ the dispensing record and the CRFs agree concerning the identification of subjects receiving the test article;
- ✔ the test articles have not expired or have not been recalled by the sponsor; and
- ✔ all test articles returned by subjects are accurately counted and recorded.

The CRA will need access to the test article storage area during site visits to verify the status of the test article supply. While some site-specific SOPs may not allow CRAs into the drug storage area due to confidentiality issues, the CRA must be able to document that the test article is being stored according to the specific Protocol requirements.

B. The Drug Accountability Record

The PI or designee (typically the CRC) is responsible for dispensing and accounting for the test articles and for exercising GCPs in monitoring these supplies. The dispensing and return of test articles must be entered promptly on the Drug Accountability Record. The sponsor or CRO/SMO should supply this form.

Under no circumstances should the PI supply test articles to other investigators, study sites, or to non-study patients, nor should she/he allow the test articles to be used in a manner other than as directed by the Protocol.

Test articles returned by a subject must be entered on the Drug Accountability Record and set aside for return to the sponsor. ***They must not be re-dispensed to anyone—not even to the same subject.***

At the termination of the study, or at the request of the sponsor or CRO/SMO, the CRA will collect all remaining supplies, and the final Drug Accountability Record will be completed. These will then be delivered to the sponsor. A copy of the Drug Accountability Record should be retained in the Study File. All manifests documenting shipments of test articles should be retained as well, along with copies of a Returned Goods Form, if provided.

During site visits, the CRA will typically be required to make a 100% source document verification of all dispensing and return data that appear on the Drug Accountability Record.

C. Case Report Forms

Case Report Forms provide for the orderly transfer of data from the study site to the CRO/SMO or sponsor. CRFs are preprinted pages that allow the investigator or CRC to write in appropriate data regarding demography, efficacy, safety, medication use, and other aspects of the study. Therefore, it is of the utmost importance that data are entered properly and accurately on the CRF.

The following guidelines should be followed when completing CRFs:

- ✔ All entries should be completed legibly using a ***black*** ballpoint pen or a typewriter.
- ✔ Only acceptable medical terminology and standard abbreviations should be used.
- ✔ Care should be taken not to write on any other forms placed on top of the CRF (written impressions will transfer to all copies of the carbonless forms underneath).
- ✔ Time should be entered as military time (a conversion chart is provided in Appendix IV, Part A).
- ✔ Numbers should be used to indicate dates (e.g., 10-01-99, not Oct. 1, 1999), unless otherwise specified.
- ✔ When completing the CRF and other documents pertaining to the study, there must be a response to every question so that the study data is complete for later entry into the sponsor's database. Blank spaces, question marks, or zeros should not be used for unknown quantities. Instead, in cases where the data requested are unavailable, the following abbreviations can be used:

N/Av—not available. This abbreviation is used when the data are not available because they are not retrievable, the subject cannot remember, or the data were lost.

N/A—not applicable. This should be used for data that do not apply to the subject. For example, if the CRF calls for the date of the last pelvic examination and the subject is a man, "N/A" should be entered.

N/D—not done. This is entered when the data were not obtained. For example, if a vital sign such as blood pressure was requested but was not measured, enter "N/D."

Note: Sponsors may provide the site with a list of acceptable, study-specific abbreviations.

- ✔ Each item should be answered or checked individually rather than by using vertical lines or "ditto" marks to indicate a series of identical answers. The CRA needs to verify that the item was addressed.

Any discrepancies or missing data noted by the CRC or the CRA during site visits should be resolved before the data are forwarded to the CRO/SMO or sponsor.

Corrections should be made in the following manner:

- ✔ The original entry should be lined out with a single line drawn through the error so it remains legible (not erased, written over, or covered up by correction fluid).
- ✔ The correction should be entered in ink and initialed and dated by the person making the correction.

Only the PI or the designee listed on the Authorized Representative Signature Record (Site Signature Log) may enter corrections on original CRFs.

The CRA is responsible for inspecting every page of data before the CRFs are sent to the sponsor or CRO/SMO. This inspection generally consists of a complete audit of all laboratory test values, the signed Informed Consent Forms, and critical non-laboratory data (e.g., inclusion and exclusion criteria) against the source documents. In addition, a spot-check of at least 20% of all other data is typically performed. The CRA will also check for legibility, completeness, and consistency.

Before a review of the CRFs is conducted, the CRC should be sure that all forms are available, completed, and signed for the CRA's inspection within two weeks of study subject contact, and provide explanations for missing data (e.g., delayed laboratory reports).

The CRC is responsible for making sure that all CRFs have been dated and signed by either the PI or an associate named on Form FDA 1572. If an associate signs CRFs, the investigator must provide signed documentation that affirms his/her review and approval of the CRFs before their release to the CRO/SMO or sponsor.

D. Authorized Representative Signature Record (Site Signature Log)

This record is maintained at each clinical study site to document the full name, handwritten signature, and initials of the PI and the representatives who are authorized to complete and/or make changes to the CRFs. A copy of this document should be maintained in the on-site investigator Study File.

If newly authorized representatives are assigned during the study, a new Authorized Representative Signature Record will be initiated for those signatures. This information should also be forwarded to the sponsor.

E. The Adverse Experience Form

An adverse experience can be broadly defined as a medical complaint, change, or possible side effect of any degree of severity that may or may not be attributed to the test article. The reporting of adverse experiences is an extremely important part of conducting a clinical study, and the CRC plays a vital role in the proper and timely reporting of adverse experiences and their follow-ups.

The Adverse Experience Form included in the CRF packet provided by the sponsor is used to record events that the investigator regards as adverse experiences (see Appendix II, Form 3). The investigator has the final decision regarding what is to be reported on the Adverse Experience Form, but the CRA may question the investigator about including unreported complaints found in the source documents or reported in other documents included in the CRF. If there is a question regarding where any study subject information is to be reported, the CRA should be contacted.

Some sponsors list concomitant medications on Adverse Experience Forms because the event may be related to the concomitant medication and not the test drug.

Serious adverse experiences must be reported promptly to the sponsor. A serious adverse event is defined as any adverse drug experience occurring at any dose that results in any of the following outcomes: death; a life-threatening adverse drug experience; in-patient hospitalization or prolongation of existing hospitalization; a persistent or significant disability/incapacity; or a congenital anomaly/birth defect. An important medical event that may not result in death, be life-threatening, or require hospitalization may be considered a serious adverse drug experience when, based upon appropriate medical judgement, the event may jeopardize the patient and may require medical or surgical intervention to prevent one of the outcomes listed in this definition.

An adverse experience can meet the definition of serious adverse event and, ultimately, not be related to the study drug. In the event of a serious adverse experience, the sponsor has ten working days during which to file a Drug Experience Report via the MedWatch Form (see Appendix II, Form 4) to the FDA. The MedWatch Form is part of an overall program designed to streamline the reporting of adverse events and product defects associated with medications, devices, and nutritional products. These reports are essential to FDA drug surveillance efforts. MedWatch Forms can be obtained from the sponsor or directly from the FDA (1-800-FDA-1088). To complete the MedWatch Form, the sponsor will need the following information from the CRC:

- ✔ subject's initials, study number, age, and sex;
- ✔ adverse experience reported, all relevant tests and laboratory data, and the status of the reaction to date;
- ✔ dates of test article administration, from the first dose to the time of adverse experience;
- ✔ whether or not the test article was discontinued and, if so, whether the reaction stopped;
- ✔ whether or not test article administration was resumed and, if so, whether the reaction reappeared;
- ✔ whether or not the investigator believed the experience was related to the test article; and
- ✔ any other factors the investigator considers to be relevant (e.g., concomitant drugs, history, and/or diagnosis).

The PI is also obligated to report in writing serious adverse experiences to the IRB within **ten working days.** Some IRBs will require a written report within 24 to 48 hrs. Make sure you review the IRB's reporting requirements.

Any abnormal laboratory values, abnormal clinical findings, or adverse experiences that the PI considers to be clinically significant must be medically managed until resolved. Follow-up information should be transmitted regularly to the CRA.

Note: If a patient dies while participating in a clinical trial, a copy of the death certificate may be required. Therefore, it is wise to obtain and have one in your files.

F. Specimen Handling and Submission Forms

The collection of biologic specimens for shipment to a reference-testing laboratory will be required by most Protocols. Common specimens include plasma, serum, urine, and tissue. The Protocol will provide specific methods for their collection, preservation, and shipping. The sponsor will supply instructions for the identification and storage of these specimens as well as the necessary submission forms to be completed at the time of shipment.

It is the responsibility of the investigator and CRC to see that specimens are collected and stored as required by the Protocol. Labels and forms identifying the subject, type of specimens collected, site of collection, and date and time of collection are to be completed at the time of the procedure by the investigator or designee.

A complete review of specimen collection and handling procedures usually takes place with the investigator and staff before initiation of the study. In the event that the sponsor is using a central laboratory, the central laboratory will provide the site's staff with a manual that is study-specific, including instructions for specimen collection and shipment.

G. The Subject Enrollment Form

The Subject Enrollment Form serves many purposes. It provides a checklist of the CRF documents to be completed for each visit and serves as a record of those components of the CRF that have been completed and submitted to the CRO/SMO or CRA. In addition, the Subject Enrollment Form is an aid in scheduling subject visits.

H. The Site Visit Log (Monitor Log)

The Site Visit Log (Monitor Log) is used to record visits made to the study site by the sponsor's authorized representative, usually the CRA or a CRO/SMO representative. At each visit, the visitor will sign the Site Visit Log. This log must be kept in the Study File. Should the FDA audit the study, the FDA inspector will want to review the Site Visit Log to determine the frequency of site visits made by the sponsor and the CRO/SMO, if applicable.

I. The Telephone Log

The Telephone Log is used to record all telephone contacts pertaining to the study. In addition to the date and time, the items discussed, action taken, and follow-up required should also be included. In some instances, the sponsor will not require a site visit follow-up letter. In these cases, the CRA may contact the site after the visit to discuss the visit. Record this conversation in the Telephone Log; some CRAs will date and initial this telephone conversation as accurate during their next monitoring visit. As with the Site Visit Log, the Telephone Log should also be kept in the Study File. A sample Telephone Conversation Record is provided in Appendix II (Form 5).

J. Delegation of Authority Log

The Delegation of Authority Log is usually required for the purpose of identifying what study-specific tasks have been delegated by the PI. This is an ICH requirement.

VIII. The Pre-Study Site Visit

Before test articles can be sent to the study site and subject enrollment can begin, an initial site visit is conducted by a representative of the sponsor and the CRO/SMO, if applicable. This is called the pre-study site visit.

Site Initiation Visit

The purpose of this visit is to give the sponsor's representative the opportunity to review the study documents, particularly the *Investigator Brochure*, Protocol, and CRFs, with the PI and staff. The *Investigator Brochure* is a document that summarizes everything known about the test article to date. One of its purposes is to inform investigators of the pharmacologic events and side effects that have been observed in past studies. The Protocol gives specific instructions for the conduct of the study. During the pre-study site visit, the sponsor's representative will discuss the Protocol and its CRFs page by page. In addition, the sponsor's representative will outline study procedures; inspect the facilities; and verify safe, secure, and appropriate storage of the test article. A representative of the CRO/SMO involved in the study may also be a participant in this review.

At the conclusion of this visit, if everything has met the approval of the sponsor, authorization to initiate the study may be given.

IX. Recruiting and Enrolling Subjects

The first step in initiating the actual clinical study enrollment is to obtain a list of potential subjects. This can be accomplished by examining subject listings from various sources and by publicizing the study. Examples are given below.

A. Prospective Subject Groups

Lists of prospective subjects can be obtained from the following:

- ✔ charts in physicians' offices;
- ✔ computer printouts by diagnosis from hospital clinic records;
- ✔ support groups;
- ✔ group screening records;
- ✔ clinical laboratory screening records; and
- ✔ referrals from physicians, nurses, coordinators, and other specialty areas (e.g., radiology, laboratory), with a referral fee offered.

B. Methods of Publicizing the Study

A clinical study can be advertised or publicized in a variety of ways. However, IRB and sponsor approval are required in advance for newspaper, radio, and television advertising and for any signs that are to be posted in public places, such as waiting rooms (see Appendix II, Form 6 for a sample sign). The following are several methods by which patients may learn about the availability of a clinical study:

- ✔ advertising (newspaper, radio, television);
- ✔ signs/posters for waiting areas;
- ✔ quarterly newsletter to subjects;
- ✔ billing statements;
- ✔ Kiwanis, Lions Clubs, Rotary Clubs, or other community groups;

- ✔ local churches and daycare centers;
- ✔ subject information in waiting rooms/examination areas;
- ✔ health fairs;
- ✔ local colleges and universities; and
- ✔ notices to local support groups for specific diseases.

Several states, including Florida, have added laws for advertisements by health care providers of free or discounted services. These laws require that specific language be included in study-related advertisement. To see if your state has enacted a law similar to the State of Florida, contact the appropriate agency in your state from which you may obtain this information. The following is the Florida Statute:

Florida Statutes Chapter 455.664

Advertisement by a Health Care Provider of Free or Discounted Services; Required Statement

In any advertisement for a free, discounted fee, or reduced fee service, examination, or treatment by a health care provider licensed under chapter 458, chapter 459, chapter 460, chapter 461, chapter 462, chapter 463, chapter 464, chapter 466, or chapter 486, the following statement shall appear in capital letters clearly distinguishable from the rest of the text:

THE PATIENT AND ANY OTHER PERSON RESPONSIBLE FOR PAYMENT HAS A RIGHT TO REFUSE TO PAY, CANCEL PAYMENT, OR BE REIMBURSED FOR PAYMENT FOR ANY OTHER SERVICE, EXAMINATION, OR TREATMENT THAT IS PERFORMED AS A RESULT OF AND WITHIN 72 HOURS OF RESPONDING TO THE ADVERTISEMENT FOR THE FREE, DISCOUNTED FEE, OR REDUCED FEE SERVICE, EXAMINATION, OR TREATMENT.

However, the required statement shall not be necessary as an accompaniment to an advertisement of a licensed health care provider defined by this section if the advertisement appears in a classified directory, the primary purpose of which is to provide products and services at free, reduced, or discounted prices to consumers and in which the statement prominently appears in at least one place.

Added by Laws 1997, c.97-261,x 81, eff. July 1, 1997

However, if the advertisement does not state "free" or "at no cost," the statement is not required. Check with your IRB for further clarification.

C. Presenting the Study to the Subject

Once a list of potential subjects has been obtained, the CRC must screen the subjects for acceptability and introduce them to the details of the study. This can be accomplished by telephone or in person.

When first contacting or meeting a potential study subject, the CRC should give his/her name and affiliation with the clinic or physician conducting the study. The reasons for conducting the current study should be explained, and an investigational drug study should be defined, if appropriate. That is, the subject should understand that the drug has been previously used in humans, but additional human data are needed before the drug may be introduced to the marketplace.

The subject is then screened according to the inclusion/exclusion criteria set forth in the study Protocol. If the subject meets the criteria, the CRC can outline the details of the study more fully.

The potential benefits and possible side effects of the study drug should be discussed. The subject should be made fully aware of the time commitment required, the need to administer the test article as instructed, the need to follow the Protocol, and the importance of compliance. The length of the study, the number of on-site visits needed, and the procedures that will be performed at each visit should be fully outlined.

If the subject has been contacted by telephone, an appointment should be made for him/her to come in to review and sign the Informed Consent Form. If the subject has been contacted in person, informed consent can be obtained at this visit. The subject must review the Informed Consent Form and then be encouraged to ask any questions regarding the study and informed consent. The subject then signs the form with the CRC acting as witness. (A duly authorized representative may act on behalf of the subject.) The PI's signature will also be obtained on this form.

Once informed consent has been obtained, the subject's next appointment (typically the baseline screening visit) should be scheduled. The subject should be given the CRC's name and phone number. The subject also receives a copy of the Informed Consent Form, any study overview, and a confirmation letter (see Appendix II, Form 8), if appropriate.

It is the responsibility of the CRC and investigator to ensure that a clinical study has been completed with the designated number of clinically acceptable subjects and that these subjects report for every scheduled visit. Therefore, when first presenting a study to a potential subject, it is important to point out the benefits of study participation. These might include the following key points:

- ✔ preferential treatment, including convenient appointment times, no delays getting in for a scheduled appointment, ample appointment time, confirmation calls for next visit, follow-up calls and letters, transportation provided if needed, personalized appraisal of laboratory test results, and thorough health care;
- ✔ free medical and laboratory tests (e.g., electrocardiogram, chest radiographs, and blood and urine analyses);

- ✔ possible monetary compensation for travel expenses;
- ✔ opportunity to participate in innovative drug research;
- ✔ opportunity to possibly help others; and
- ✔ opportunity to increase understanding about disease processes.

To ensure fair balance, the subject must also be made aware of the potential risks involved in participating in the study.

On occasion, a subject will require special instructions about a particular procedure or test. The following box provides an example of one such special information directive.

Instructions for 24-Hour Urine Collection

The same amount of liquids should be consumed during the collection period as are normally consumed. No alcoholic or caffeine-containing beverages (coffee, tea, cola) may be consumed during this period.

The collection starts **after** you empty your bladder in the morning. *(The urine voided at this time is not included in the collection.)* Write the time and date on the label of the specimen container. Collect all urine voided for the next 24 hours in this one container, including the first specimen voided the following morning.

The specimen container should be refrigerated during the collection period. Each voiding should be added to the container as soon as possible.

Proper collection and refrigeration of the urine specimen is very important for accurate test results. Please follow the above directions and call me if you have any questions.

[Name]
Clinical Research Coordinator

[Phone number]

X. Conducting the Study and Keeping Records

A. Study Activities

All the work performed up to this point should facilitate the accurate and timely transfer of study data to the CRFs. However, while the clinical study is being conducted, the CRC must ensure that clinical findings and laboratory data are properly recorded. Toward this end, the CRC should:

- ✔ organize and label specimen containers;
- ✔ prepare all laboratory requisition sheets;
- ✔ prepare subject workfolders before each visit;
- ✔ oversee the receipt and handling of test articles and laboratory specimens;
- ✔ maintain all files, records, and logs; and
- ✔ prepare for site visits.

For a study to be successful, *the protocol must be followed and subject numbers maintained.* Some common pitfalls to avoid are:

Lack of Timely Treatment. This may occur, leading to subject withdrawal if the physician's office staff is not alerted to the status of a study subject. To avoid this problem, the names of study participants should be highlighted in the physician's appointment book.

Laboratory Specimens Not Analyzed. Laboratory specimens must be processed within a certain period of time to meet the requirements of the laboratory or the Protocol. If specimens are not processed in a timely manner, the Protocol may be violated or important laboratory data may have to be excluded. When possible, a back-up sample should be retained.

Subject Absence. Subjects must show up for the required number of office visits, as dictated by the Protocol. To minimize subject absence, the CRC should telephone or mail a postcard to remind the subject of the upcoming appointment.

B. Subjects Lost to Follow-Up

If a subject is continually absent for appointments, the CRC should try to contact him/her by phone or certified letter to determine any scheduling conflicts and, if possible, reschedule the missed appointments. If attempts at follow-up using the Informed Consent Form and/or Letters of Agreement fail to convince the subject to report for scheduled appointments, the subject must be withdrawn from the study. If this is the case, every attempt should be made to have the test articles returned to the study site. The dates the subject was contacted and the type of contact used should be recorded in the study documentation.

C. Withdrawing a Subject

A subject may be withdrawn (terminated) from a study for several reasons, including failure to follow the Protocol or having an adverse reaction to the test article. A subject may also voluntarily withdraw. In the event a subject is withdrawn, a final visit should be scheduled, and the sponsor should be notified of the subject's withdrawal. The subject's randomization code should not be broken at this time except in the case of an emergency. Many study Protocols require follow-up to be continued for a specified period of time after a subject is withdrawn, regardless of the reason for withdrawal.

D. Site Visits During the Study

The study site will be visited several times by the CRA or representative of the CRO/SMO. The complexity of the study or the number of subjects enrolled will dictate the number of visits. In addition, the CRA will monitor the study site by telephone. At each telephone contact or site visit, the CRA will confer with the CRC and the PI regarding the progress of the study.

During a typical site visit, the CRA is required to perform the tasks that follow. A review of these items will help the CRC prepare for the CRA's site visit.

Sign Site Visit Log (Monitoring Log). All of the sponsor's personnel must sign the log each day they visit the study site.

Inspect original, signed Informed Consent Forms. These must be available for all subjects enrolled since the CRA's previous site visit.

Inspect CRFs. Case Report Forms must be completed legibly using a black ballpoint pen or typewriter. They must be available for review within two weeks after the subject completes the study.

Inspect Applicable Source Documents. The CRA will check for:

- ✔ properly completed entries;
- ✔ consistency between CRFs. All laboratory data and at least 20 percent of all other entries will be checked against source documents. Most studies involve a 100 percent quality control check;

- ✔ conformity with the protocol; and
- ✔ properly signed and dated CRFs. (These must be signed by the PI or the authorized designee.)

Inspect Test Article Storage Area. The CRA will require access to this area to ensure that:

- ✔ adequate supplies are on hand;
- ✔ test articles are being stored properly;
- ✔ the Drug Accountability Record is being properly maintained;
- ✔ there is agreement amongst a physical count of the test articles received versus the number of test articles dispensed (as recorded on the Drug Accountability Record and CRFs) versus the number of test articles on hand;
- ✔ there is agreement between the dispensing record and the CRFs regarding the identification of subjects receiving the test article; and
- ✔ all test articles returned by subjects have been accurately counted and recorded on the Drug Accountability Record.

Inspect Bioanalytical and Specimen Storage Area. The CRA will need access to this area to ensure that samples collected since the last visit have been properly collected, handled, and stored.

E. Closing the Study and Retaining Records

At the termination of the study or at the request of the sponsor, the CRA or representative of the CRO/SMO will pick up all remaining supplies and the final Drug Accountability Record for transmittal to the sponsor. A copy of the Drug Accountability Record will be retained in the Study File.

Study Close-Out Checklist. At the close of a study, the CRC is required to:

- ✔ schedule a close-out meeting with the sponsor;
- ✔ notify the IRB of close-out and submit a final report;
- ✔ submit a final report to the sponsor;
- ✔ submit miscellaneous charges to the sponsor;
- ✔ return or discard laboratory supplies;
- ✔ bind and store CRFs;
- ✔ return or discard unused CRFs;
- ✔ label and ship frozen specimens;
- ✔ complete and store audit file;

- ✔ request randomization information; and
- ✔ track disbursements and reconcile study finances with the sponsor.

IRB Final Report. The IRB Final Report is submitted after all the subjects have completed the study. This is sent to the IRB, with a copy sent to the sponsor.

Record Retention. Federal law requires that the PI retain all study records for two years following the NDA approval date. The sponsor may request longer retention periods. If the application is not approved, or no NDA is submitted, the PI must retain all study records for at least two years after the FDA has been notified that all clinical investigations of this indication have been discontinued. The ICH guidelines recommend that the records be retained for 15 years. When archiving study records, clearly label the following:

- ✔ Protocol number;
- ✔ Sponsor name;
- ✔ investigator name;
- ✔ investigator address; and
- ✔ study close-out date.

The PI must obtain the written consent of the sponsor before disposing of any study records.

Subject Follow-up. Subjects appreciate receiving correspondence after a study has been completed. You may wish to send them a letter describing the drug group in which they participated, you may provide a self-assessment form so they may rate their satisfaction with the conduct of the study, or you may wish to thank them for their participation.

XI. Preparing for an FDA Audit

A. FDA Inspections of Clinical Trials

The FDA Bioresearch Monitoring Program involves site visits to PIs, research sponsors, CROs/SMOs, IRBs, and non-clinical (animal) laboratories. All FDA product areas such as drugs, biologics, medical devices, radiological products, foods, and veterinary drugs are involved in the Bioresearch Monitoring Program. The program procedures are dependent upon the product type. The objective of all FDA inspections is ensuring the integrity and quality of data and information submitted to the FDA, as well as the protection of human research subjects.

The FDA carries out three distinct types of clinical trial inspections:

✔ study-oriented inspections;

✔ PI-oriented inspections; and

✔ bioequivalence study inspections.

Study-Oriented Inspections

FDA field offices conduct study-oriented inspections based on assignments developed by headquarters staff. Assignments are based almost exclusively on studies that are important to product evaluation, such as NDAs and product license applications pending before the FDA.

When a PI who has participated in the study being examined is selected for an inspection, the FDA investigator from the FDA district office will contact the PI to schedule a mutually acceptable time for the inspection.

Upon arrival at the site, the FDA investigator will show FDA credentials, consisting of a photo identification, and present a Notice of Inspection Form to the PI. Food and Drug Administration credentials let the PI know that the FDA investigator is a duly authorized representative of the FDA.

If, during the course of a FDA inspection, a PI has any questions that the FDA investigator has not answered, either the director of the district office or the center that initiated the inspection may be contacted. The name and telephone number of the district director and the specific center contact person are available from the FDA investigator.

B. Types of FDA Inspections

There are two main types of FDA inspections that you are likely to encounter: a routine inspection and a for-cause inspection.

A *routine* inspection is conducted for those studies that are crucial to a product's evaluation and approval. For example, the studies may be pivotal to an NDA pending before the FDA. As the name would imply, a *for-cause* inspection is conducted when the FDA has a specific reason (cause) for inspecting a study site. The reason might be one of the following: (1) an investigator has participated in a large number of studies; (2) an investigator has done work outside his/her specialty; (3) the safety or efficacy results of an investigator are inconsistent with those of others conducting the same study; or (4) the investigator claims too many subjects with a specific disease or indication compared with the low numbers associated with his/her practice. Numerous other reasons determined by the FDA may trigger a for-cause inspection.

Before an inspection, the FDA will contact the PI, usually through a letter, to schedule the date of inspection and to determine the specific material to be audited. After receiving notice of a pending inspection, the PI should immediately contact the CRA because the sponsor may wish to meet with site personnel prior to the FDA inspection.

The FDA inspector will present FDA credentials and a completed Form FDA 482, Notice of Inspection. (This form can be located on the FDA Web site.) The CRC should reserve a private area for the inspector, such as a conference room, and organize all appropriate source documents for the inspection. The inspector should have access to all, and only, the study data that was specifically requested, nothing more and nothing less. In addition, the following related information should be available:

- ✔ physicians' office records;
- ✔ hospital records;
- ✔ laboratory test results;
- ✔ subject medical history;
- ✔ subject follow-up data; and
- ✔ appointment calendar.

The inspector may also request the following information:

- ✔ the degree of delegation of authority by the PI;
- ✔ how and where data were recorded;

- ✔ how test article accountability was administered and maintained;
- ✔ how the sponsor communicated with the PI and evaluated the study program; and
- ✔ certification of service/calibration for study-related equipment.

C. Parts of the Investigation

The investigation consists of two basic parts. The first is determining the facts surrounding the conduct of the study, including:

- ✔ who did what during the conduct of the clinical trial;
- ✔ the degree of delegation of authority performed by the PI— whether or not they were properly supervised and had the experience to do the assigned task;
- ✔ where specific aspects of the study were performed;
- ✔ if they are listed in the FDA 1572;
- ✔ how and where the data were recorded;
- ✔ how test article accountability was dispensed and maintained;
- ✔ how the CRA communicated with the clinical investigator; and
- ✔ how the CRA monitor evaluated the progress of the clinical trial.

Second, the clinical trial data is audited. The FDA investigator compares the data submitted to the FDA and/or the sponsor with all available records that might support the data. These records may come from the physician's office, hospital, nursing home, central and/or local laboratories, and other sources. The FDA may also examine patient records that predate the study to determine whether or not the medical condition being studied was, in fact, properly diagnosed and whether or not an interfering medication had possibly been given before the study began. The FDA investigator may also review records covering a reasonable period after completion of the clinical trial to determine if there was proper follow-up, and if all signs and symptoms reasonably attributable to the product's use had been reported.

1. Principal Investigator-Oriented Inspections

A PI-oriented inspection may be initiated because a PI conducted a pivotal study that merits in-depth examination because of its singular importance in product approval or its effect on medical practice. An inspection may also be initiated because representatives of the sponsor have reported to the FDA that they are having some difficulty getting case reports from the PI, or that the FDA has some other concern with the PI's work.

In addition, the FDA may initiate an inspection if a subject in a study complains about the Protocol or human subject right's violations.

Principal investigator-oriented inspections may also be initiated because:

- ✔ the PI participated in a large number of studies;
- ✔ the PI enrolled a significantly large number of subjects into a Protocol;
- ✔ the PI participated in clinical trials outside of his/her specialty areas;
- ✔ the safety or effectiveness findings were inconsistent with those of other PIs studying the same test article;
- ✔ too many subjects with a specific disease given the locale of the investigations were claimed; and/or
- ✔ the laboratory results were outside the range of expected biological variation.

Once the FDA determines that a PI-oriented inspection should be conducted, the procedures are essentially the same as in the study-oriented inspection, except that the data audit goes into greater depth, covers more case reports, and may cover more than one study. If the investigator has repeatedly or deliberately violated FDA regulations or has submitted false information to the sponsor in a required report, the FDA will initiate actions that may ultimately decide that the PI is not to receive investigational products in the future.

2. Bioequivalence Study Inspections

A bioequivalence study inspection differs from the other inspections in that it requires participation by a FDA chemist or a PI who is knowledgeable about analytical evaluations.

D. Common FDA Inspection Findings

At the end of an investigation, the FDA investigator will conduct an "exit interview" with the clinical investigator. During this interview, the FDA investigator will discuss the findings from the inspection with the PI. The FDA investigator will clarify any misunderstandings that might exist, and may issue a written Form FDA 483 (Notice of Observations) to the PI. Following the inspection, the FDA field investigator will prepare a written report and submit it to headquarters for evaluation.

Once the report has been evaluated, the FDA headquarters usually issues a letter to the PI. The letter is generally one of three types:

> **Notice that no significant deviations from the regulations were observed.**
> This letter does not require any response from the PI.
>
> **Informational letter** that identifies deviations from the federal code of regulations and GCP. This letter may or may not require a response from the PI. If a response is requested, the letter will describe what is necessary and provide a contact person for clarification or further questions.

Warning letter that identifies deviations from the code of federal regulations requiring prompt correction by the PI. This letter will provide the PI with a contact person for clarification and questions. In these cases, the FDA may inform both the sponsor and the reviewing IRB of the deficiencies. The FDA may also inform the sponsor if the PI's procedural deficiencies indicate ineffective monitoring by the sponsor. In addition to issuing these warning letters, the FDA may take other courses of action, which may include regulatory, legal, and/or administrative sanctions.

Office for Human Research Protections

In June 2000, the FDA announced the creation of a new office at the Department of Health and Human Services, the Office for Human Research Protections (OHRP), to lead efforts for protecting human subjects in biomedical and behavioral research. This office replaces the Office for Protection from Research Risks (OPRR), which was part of the National Institutes of Health. The OPRR had authority over NIH-funded research.

According to the HHS Assistant Secretary, this new office will have increased resources and broader responsibility to ensure that patients taking part in research are better protected and fully informed. Everyone in the research community must share the responsibility for protecting research subjects. Medical research today is exploding with opportunity; however, to achieve the benefits of that research, we need a solid foundation of thoughtfully designed and thoroughly executed research-patient protection. This new OHRP will work with HHS agencies, research institutions, and sponsors of research to ensure that this foundation is in place and working productively.

For additional information, a copy of the FDA *Compliance Program Guidance Manual* for Clinical Investigator Inspections (Program 7348.811), the document issued by the FDA investigator to conduct the inspection, is available by writing to:

Freedom of Information Staff (HFI-30)
Federal Food and Drug Administration
5600 Fishers Lane
Rockville, MD 20857

In addition, see FDA information sheet, "Clinical Investigator Regulatory Sanctions," available at the FDA Web site.

E. Common Problems Identified in Audits of Clinical Trials

Audit Problem Areas	Items for Caution
Protocol	Under what Protocol date did the site initiate the study? If there are several versions, is the initiation version at the site and is it documented that this is the appropriate Protocol version? In addition, are all versions of the Protocol in the Investigator File?
Protocol Signature Page	Are corresponding completed signature pages provided for all versions of the Protocol?
Investigator Brochure (IB)	What version of the IB was provided to the site? Was the IB revised during the clinical trial? Is there documentation showing that the IB was received by the site (e.g., a memo to the Investigator File stating when the document was sent or a monitoring report listing the document as being at the site)? Where is the IB kept? If it is not kept in the Investigator File, is there a memo stating where at the site the IB can be located?
Form FDA 1572	Has the form been correctly completed, signed, and dated? If there is more than one form with the same signature/date, which form is correct? Has section #8 been appropriately completed? If Form FDA 1572 has been revised, was a copy sent to the IRB as well as the sponsor?
Curriculum Vitae (CV)	Does the site have a CV for each person listed on FDA Form 1572? Does the site have a current medical license for the PI (some sponsors require a copy of the medical license for the sub-investigator(s) as well as the PI)? Are all CVs current within two years? Can this be easily determined by review of the CV? Does the CV document the PI's affiliation with the site conducting the clinical trial?
Confidentiality Agreement	Was the Confidentiality Agreement obtained prior to the release of all study-related documentation (e.g., protocol, IB, Financial Disclosure form)?
Informed Consent Forms	Is there a copy of the IRB approved consent on file? Does the IRB documentation indicate that it is approving the consent? Has the informed consent been revised? Was the revised consent forwarded to the IRB? Is there documentation indicating that revised consent was approved? Have the patients participating in the Protocol signed the correct version of the informed consent? Are all sections of the informed consent signed and dated appropriately? If the site has foreign-speaking patients, is there a copy of the translated IRB approved consent? Has a certified translator completed the translation? Is there a back translation showing exactly what information is being presented to the patient? If the study has enrolled minors, is guardian consent obtained? If appropriate, is there assent by the minor participating in the Protocol? If there is another individual signing for the patient (e.g., daughter signs for father or wife signs for husband), is the individual the legal representative of the subject? If this is an in-hospital clinical trial, has the hospital's requirements for a legal representative been met? Has the reason there is another individual signing the informed consent for the patient been documented? Have all informed consent documents been signed prior to the initiation of any study-related screening procedures (e.g., medication wash-out periods, fasting for screening laboratory tests)?
Serious Adverse Events	Have all serious adverse events been reported to the sponsor, CRO/SMO, and IRB within the specified time frames? Does the documentation provide a complete picture of the event? Has the condition been followed until resolution? In the event the patient has died, is there a copy of the death certificate in the source documents?
Institutional Review Board (IRB)	Does the site have documentation that the IRB is appropriately formed (e.g., membership list or assurance number)? Can all IRB approvals be easily tracked to the corresponding study document? Are there periodic reviews being appropriately submitted to the IRB and in a timely fashion? Is there documentation in the Investigator File from the IRB in response to the reviews? Are study summaries sent to the IRB upon study completion? Are all serious adverse events being appropriately reported? Are significant Protocol departures being reported to the IRB? Is there documentation in the Investigator File from the IRB in response to the reports?

Audit Problem Areas	Items for Caution
Site Visit Follow-up Documentation	Are issues identified during a site-monitoring visit followed up in subsequent visits? Is there documentation to indicate that identified issues have been appropriately resolved?
Laboratory Documentation	Does the site have the appropriate laboratory certification (e.g., CLIA, CAP, and state license where applicable)? Does the certification cover the duration of the study? Are there normal laboratory value ranges available? If a central laboratory is being used for the study and there were Protocol samples sent to a local laboratory, is the appropriate certification for both the central and the local laboratories available in the Investigator File?
Investigational Supplies	Can all investigational supplies be accounted for? Are all study-related supplies being appropriately stored (e.g., if test articles need refrigeration, are they stored in a refrigerator with a temperature monitor and log? Is the temperature log current? If the test article is a controlled substance, have the appropriate guidelines for storage and dispensing been performed?)
Site Equipment Specific to the Protocol	Has the equipment being used been serviced recently? Are the service logs available? If calibration is required, has the calibration been performed prior to the initiation of any study-related procedures? Is there documentation of the calibration?
Correspondence	Is there sufficient correspondence between the site and the sponsor, the site and the IRB, and the site and CRO and/or SMO to demonstrate ongoing communication for the duration of the study? Have all required documentation been forwarded to the IRB, including periodic reports, annual reviews, and study summaries?
Research Charts (Source Documents)	Does the subject's research chart capture sufficient information to demonstrate participation in the clinical trial? Does the information transcribed on the CRFs substantiate the research chart? Are adverse events and serious adverse events recorded in the subject's research chart?

XII. Clinical Research: Potential Liability

A. Fraud

In clinical research, fraud is the deliberate reporting of false or misleading data and/or the withholding of reportable data, with the intent to mislead the sponsor. Once the sponsor has discovered the fraud, the sponsor should report the PI's misconduct to the IRB and the FDA for further investigation.

1. Types of Fraud

The definition of fraud includes:

- ✔ the misrepresentation made by the PI and/or her/his representative by words or through conduct;
- ✔ the misrepresentation made by the PI and/or her/his representative with the knowledge that the representation was not true;
- ✔ representation was made with an intention that should cause the sponsor to act on it; and
- ✔ the sponsor's use of the clinical data collected by this PI and/or her/his representative and damage suffered by use of this data.

2. Who Can Commit Fraud?

Anyone associated with the collection, transcribing, reporting, or monitoring of clinical trial data. This includes the:

- ✔ PI;
- ✔ sub-investigator(s);
- ✔ Clinical Research Study Coordinator;
- ✔ CRA;
- ✔ data manager; and
- ✔ study patient.

3. Methods Used by the Pharmaceutical Industry to Detect Fraud

These include:

- ✔ identifying the original source data;
- ✔ verifying the existence and accuracy of the source data;

- ✔ challenging the integrity of the data;
- ✔ tracking missing patient records;
- ✔ recognizing fabricated data that will point in the direction of inconclusive results favoring the test article;
- ✔ observing if all the documents are prepared or written by the same person;
- ✔ noting a repeated pattern of data and identifying departures from anticipated trends;
- ✔ determining if the PI and her/his staff has the expertise, capability, and equipment to perform the clinical trial;
- ✔ determining the sequence of events and considering whether or not it was physically possible for the individual(s) to perform the work in the given time frame;
- ✔ observing patient diary cards that are pristine;
- ✔ noticing that the handwriting is the same or very similar on the diary cards;
- ✔ observing that the same writing implements were used to complete patient diary cards;
- ✔ observing that the data collected during the course of the clinical trial are perfect;
- ✔ comparing laboratory data for all subjects enrolled in the clinical trial;
- ✔ observing when patients were seen at the site (e.g., Were patients seen on holidays or weekends when it was not required by the design of the clinical trial?);
- ✔ noting a lack of adverse events, especially in certain patient populations with specific disease indications;
- ✔ noting similar blood pressure results for all patients enrolled in the clinical trial; and
- ✔ comparing returned study medication to determine if there is a pattern to the way the blister packs were opened.

4. Investigator Misconduct

Principal investigator misconduct can be defined as follows:

- ✔ a flagrant, but non-deliberate violation of the Code of Federal Regulations and GCPs;
- ✔ modification of the research data to improve
  - ✔ ability to publish and
  - ✔ accountability;

- ✔ intentional violation of the code of federal regulations;
- ✔ deliberate fabrication of clinical trial results;

Principal investigator minor negligence can be defined as:

- ✔ disorganized data and/or
- ✔ sloppy record-keeping.

However, the CRC and the CRA can correct these issues via proper documentation.

B. Negligence

Principal investigator negligence can exist when some of the clinical research data are compromised. The compromised data collected by this PI must be removed from the study database and must not be included in the statistical analysis of the clinical trial.

Principal investigator gross negligence can exist when the PI has failed to exercise control over the clinical trial in question. In this instance, all the data generated by this PI is compromised and *all* of the efficacy data collected during the conduct of the clinical trial must be removed from the study database and must not be included in the statistical analysis of the clinical trial.

According to an article in *The New York Times* in May 1999, in an era of managed care the number of private physicians in research since 1990 has almost tripled, and top-enrolling physicians can earn as much as $500,000 to $1 million a year. Pharmaceutical companies and their contractors (CROs/SMOs) offer large payments to physicians and other allied health professionals to encourage them to enroll patients quickly. There are finder's fees paid to physicians for referring their patients to other physicians conducting clinical trials. In some instances, there are payments to everyone involved in the clinical trial that can assist in the recruitment of patients.

Physicians with money at stake may persuade patients to take drugs that are inappropriate or even unsafe for them due to pre-existing medical conditions. Unfortunately, clinical research fraud has become a problem in recent years, resulting in the criminal prosecution of several physicians and study coordinators.

Study patients may also commit fraud by providing false information to participate in a specific study. If patient fraud is detected, notify the sponsor immediately.

C. Legal Liability and the Clinical Research Coordinator

Criminal Liability is the violation of statutes that carry criminal penalties. These can be state or federal statutes. The Federal FDA regulates clinical research, and a violation of the regulations could result in federal and state criminal prosecution.

Actions that could result in criminal liability include:

- ✔ intentionally creating false or misleading data for submission to the FDA; and
- ✔ assisting with/covering up intentionally false or misleading data created by a PI for the FDA.

In some instances, a criminal prosecutor may go after a CRC to get the CRC to testify against the PI. ***A good rule to follow: If you are asked to do something, and you think it is wrong, don't do it!***

Civil Liability occurs when deliberate actions could result in liability for monetary damages.

A CRC could be considered the agent of the PI if the coordinator performs tasks delegated by the PI. A CRC could also be an agent of their institution (e.g., hospital, medical school, and research institute).

Any actions performed improperly during the course of a clinical trial that result in liability might cause the CRC, the PI, or the institution of employment to be liable.

A PI who does not properly supervise those to whom he/she delegates tasks, resulting in harm to the patient, may also be liable.

There are professionals who think that the one with the most money is the one who is sued. This is not true! The fact that you are not a "deep pocket" won't protect you from being named in a lawsuit!

XIII. Writing the Study Summary

A vast amount of information is collected during the course of a controlled clinical study. To transform this raw data into useful information, the sponsor will subject the data to rigorous analyses. The following are typical of the analyses and presentations prepared by the sponsor.

A. Biostatistical Report

Following the statistical analysis of the data collected during the study, comprehensive computer-generated tabulations are prepared to summarize the trends and changes that occurred. An interpretation of these statistical results is provided in the biostatistical report, which is typically written by a biostatistician.

B. Integrated Clinical Study Report

To present the results of the study to the FDA and other regulatory agencies around the world, the sponsor will prepare a report that provides a detailed description of the methods, results, and conclusions drawn from the study. This report integrates the key aspects of the biostatistical report with clinical interpretations of the study results. In this manner, a comprehensive overview of the study is prepared for future reference.

C. Articles for Publication

The results of key studies are often written up for submission to scientific journals. The PI or sub-investigators may be given the opportunity to draft the manuscript based on data contained in the Integrated Clinical Study Report. The sponsor may offer technical or editorial assistance in preparing these articles. Alternately, data from a study may be presented as an abstract or poster at a scientific meeting.

XIV. Achieving Credibility and Recognition as a Clinical Research Coordinator

Achieving credibility and recognition as a CRC in the pharmaceutical industry is not an easy undertaking. To be successful in the pharmaceutical research industry, it is essential to be seen as competent, believable, and confident by the CRAs, sponsors, and others.

Three important ways to enhance your credibility as a research professional are to continually radiate a professional image, to demonstrate your integrity, and to build good professional relationships.

Presenting a professional relationship utilizing written correspondence can be achieved in many ways.

- ✔ Follow a business letter format.

- ✔ Use letterhead and include the name of the study about which you are writing.

- ✔ Do not write personal comments or opinions in study-related correspondence.

- ✔ Remember that people you have never met may review anything you write. It may become a permanent part of a study's regulatory file.

- ✔ Use spell-check, and remember to proofread everything to ensure you are using the correct words (i.e., "principal" investigator versus "principle" investigator).

- ✔ Save personal comments for a time when you are speaking person-to-person.

Presenting a professional image utilizing voice mail can also be achieved.

- ✔ Generally, the same rules apply to voice and written correspondence.

- ✔ Remember that voice-messaging systems can be an open forum, as your message can be forwarded to multiple people.

- ✔ Take a moment and make sure you are leaving a message for the correct person at his/her extension. If you inadvertently leave a voice mail message on the wrong extension, the message may take on a "life of its own," as it is forwarded company-wide until/if the message reaches the person for whom it was intended.

- ✔ Messages should be *short* and to the point.

- ✔ Briefly indicate the reason why you are calling. This will better enable the CRA to quickly respond to your needs.

Presenting a professional image utilizing email can be achieved in many ways.

- ✔ Email messaging can be an open forum because your message, even if it is bcc (back copied) and marked "confidential," can be forwarded to multiple people.
- ✔ Make sure you are sending the email to the correct email address.
- ✔ Do not forward the following types of email messages:
    - ✔ chain emails and
    - ✔ emails using profanity.
- ✔ Do not put your favorite CRA on any kind of email mailing list. Sponsors may periodically monitor their employees' email messages.

Demonstrating your integrity can be achieved via the following ways:

- ✔ making sure you keep commitments;
- ✔ letting the CRA know if you are not going to be ready for his/her visit, not at the last minute, but in enough time so that the visit can be rescheduled;
- ✔ knowing that mistakes are made by everyone and that research is a team effort (the CRA is only doing his/her job);
- ✔ not taking on more work than you can realistically handle; and
- ✔ being careful not to speak negatively about another professional. You never know when someone who knows of the individual about whom you are speaking may be listening and repeat the conversation.

XV. Appendices

APPENDIX I: Helpful Sources

A. Names and Addresses of Clinical Research Resources

Membership Organizations

DIA (Drug Information Association)
P.O. Box 3113
Maple Glen, PA 19002
Telephone: 215-628-2288
Fax: 215-641-1229
Web address: www.diahome.org

American Society of Law, Medicine and Ethics
765 Commonwealth Avenue
Suite #1634
Boston, MA 02215
Telephone : 617-262-4990
Fax: 617-437-7596
Web address: www.aslme.org

PRIM&R (Public Responsibility in Medicine & Research)
Fourth Floor
132 Boylston Street
Boston, MA 02116
Telephone: 617-423-4112
Fax: 617-423-1185
Web address: www.aamc.org/research/primr
PRIM&R's Executive Office is also the Administrative Site for ARENA (Applied Research Ethics National Association)

ACRP (Association of Clinical Research Professionals)
1012 14th Street NW
Suite #807
Washington, DC 20005
Telephone: 202-737-8100
Fax: 202-737-8101

Coordinator Resources

MedTrials, Inc.
2777 Stemmons Freeway
Suite #975
Dallas, TX 75207
Telephone: 214-630-0288
Fax: 214-630-0289
Web address: www.medtrials.com
Develops site efficiency tools to increase productivity and optimize performance and regulatory compliance. They offer a line of performance products for investigative sites including:
- *Correspondence templates*
- *Source document templates*
- *Subject identification stickers*
- *Visit reminder cards*
- *Study day estimators*
- *Reference charts*
- *Patient and research staff educational materials*
- *Standard operating procedure templates for clinical research sites*

Publications

Ethics and Regulations of Clinical Research by Robert J. Levine
2nd Edition 1988—Paperback $25.00
Yale University Press—203-432-0960

Glossary of Lay Language Synonyms for Common Terms Used in Informed Consent Documents for Clinical Studies—A Handbook for Clinical Researchers by Deborrah Norris
1996—Paperback $39.95
Plexus Publishing, Inc.—609-654-6500

Seminars/Workshops

Contact the following organizations directly for information:

WIRB (Western Institutional Review Board)
Telephone: 360-754-9248 *(for their one-day coordinator class)*
             360-943-1410
Email: wirb@wirb.com

DIA (Drug Information Association)
Telephone: 215-628-2288

MED EXEC International
Telephone: 800-507-5277

PRIM&R/ARENA
Telephone: 617-423-4112

American Society of Law, Medicine and Ethics
Telephone: 617-262-4990

ACRP (Association of Clinical Research Professionals)
Telephone: 202-737-8100

FDA (Food and Drug Administration)
Telephone: 301-443-2894

Chesapeake Research Review, Inc.
Matthew Whalen, Ph.D.
President, Chesapeake Research Review, Inc.
9017 Red Branch Road
Suite 100
Columbia, MD 21045
Telephone: 410-884-2900

Legal Services

Pre-Paid Legal Services, Inc.
Telephone: (toll free) 877-508-7363
Web address: www.prepaidlegal.com/go/
Email: davidppl@ddbarwick.com
This 28-year-old NYSE company offers legal representation through its national network of provider legal firms. For additional information, details of coverage, or an application, contact: David or Darlene Barwick, Independent Associates, Pre-Paid Legal Services, Inc. Nationwide coverage available.

Commercial IRBs (Central IRBs)

(For a complete listing of all commercial IRBs, contact HIMA)

Health Industry Manufacturers Association (HIMA)
1200 G Street NW
Suite #400
Washington, DC 20005-3814
Telephone: 202-783-8700
Fax: 202-783-8750
Web Address: www.himanet.com/

Employment Services

MED EXEC International
100 North Brand Blvd.
Suite #306-308
Glendale, CA 91203

Telephone: 818-552-2036; (toll free) 800-507-5277
Fax: 818-552-2475
Web Address: www.medexecintl.com
Contact: Rosemary Christopher
(Placement Services for CRCs; Site Managers; Site Directors; Patient Recruitment Specialist)

CRC Certification
(Contact them directly for additional information)
ACRP (Association of Clinical Research Professionals)
Telephone: 202-737-8100

Food and Drug Administration
Associate Director for Science and Medical Affairs
Center for Drug Evaluation Research
5600 Fishers Lane
Rockville, MD 20857
Telephone: 301-443-2894
Fax: 301-443-2763

B. FDA Contacts for Human Subject Protection

Office of Health Affairs
Health Assessment Policy Staff
HFY-20
5600 Fishers Lane
Rockville, MD 20857
Telephone: 301-827-1685
Fax: 301-443-0232

Paul W. Goebel, Jr.: 301-827-1699

Paula Squire Waterman, M.S.: 301-827-3753

FDA home page . . . . http://www.fda.gov
FDA Information Sheets . . . . OHA Fax on demand 800-993-0098 or 301-827-3156
FDA Information Sheets . . . . http://www.fda.gov/oc/oha/toc.html
Title 21 CFR on the Web . . . . http://www.access.gpo.gov/nara/cfr
21 CRF 50.24 Consent Waiver . . . . http://www.fda.gov/opacom (more choices) (fed996)
Disqualified Investigators http://www.fda.gov/oha/list2.htm

Center for Drug Evaluation and Research
Human Subject Protection Team
HFD-343
Division of Scientific Investigations
7520 Standish Place

Rockville, MD 20855
Telephone: 301-594-1026
Fax: 301-594-1204

CDER Fax on Demand: 800-342-2722 or 301-342-2722

CDER new document list subscription. Send email message:
"SUBSCRIBE CDERNEW your mail address @ your domain" to
FDAlists@www.fda.gov
ICH guidelines for GCP May 9, 1997 FDA

C. FDA Office of Health Affairs, World Wide Web Sites of Interest for Human Subject Protection Information

While the FDA reports that there are approximately 4,000 requests per hour, it is virtually impossible to stay fully informed on all aspects related to a topic as expansive as the protection of human subjects. The following are a few FDA and federal government sites that contain information of interest to clinical research professionals, allied health care professionals, and IRBs.

FDA Home Page . . . . http://www.fda.gov/
This address is the gateway to all FDA information, with sections provided for product-specific centers (drugs, biologics, medical devices, veterinary drugs, foods, toxicological research) and FDA-wide subjects.

Information for Health Professionals
. . . . http://www.fda.gov.opacom/morechoices/moreheal.html
This site provides information targeted to health professionals.

Information Sheets . . . . http://www.fda.gov/oc/oha/toc.html
This site provides information targeted to clinical investigators and IRBs.

FDA Disqualified Investigator List http://www.fda.gov/oha/list2.htm
This site provides a cumulative list of clinical investigators who have been, or are currently, disqualified from access to investigational products or had their use of investigational products restricted.

PHS List of Investigators Subject to Administrative Action
. . . . http://silk.nih.gov/public/cbz1bje.@www.orilist.html
This site provides a list of investigators disqualified by the FDA. It includes those investigators subject to action by the HHS Office of Research Integrity.

Government Printing Office . . . . http://www.access/gpo.gov/su_docs/
This site provides any Federal register document since 1994, and any section in the entire Code of Federal Regulations. The government printing office is responsible for printing the Federal Register, Code of Federal Regulations, and Congressional Records.

Office of Protection from Research Risks . . . . http://www.nih.gov/grants/oppr/oprr.htm
Allied health care professionals involved with Department of Health and Human Services funded research are subject to regulation by this office.

The FDA encourages allied health care professionals to inform them of any additional Web sites that may be useful. They can be contacted at:

Office of Health Affairs, Health Assessment Policy Staff
Telephone: 301-827-1685
Fax: 301-443-0232

APPENDIX II: Forms Used During the Course of a Clinical Study

Form 1. THE STATEMENT OF INVESTIGATOR FORM (FDA 1572)

DEPARTMENT OF HEALTH AND HUMAN SERVICES PUBLIC HEALTH SERVICE FOOD AND DRUG ADMINISTRATION **STATEMENT OF INVESTIGATOR** *(TITLE 21, CODE OF FEDERAL REGULATIONS (CFR) PART 312)* (See instructions on reverse side.)	Form Approved: OMB No. 0910-0014. Expiration Date: September 30, 2002. *See OMB Statement on Reverse.*
	NOTE: No investigator may participate in an investigation until he/she provides the sponsor with a completed, signed Statement of Investigator, Form FDA 1572 (21 CFR 312.53(c)).

1. NAME AND ADDRESS OF INVESTIGATOR

2. EDUCATION, TRAINING, AND EXPERIENCE THAT QUALIFIES THE INVESTIGATOR AS AN EXPERT IN THE CLINICAL INVESTIGATION OF THE DRUG FOR THE USE UNDER INVESTIGATION. ONE OF THE FOLLOWING IS ATTACHED.

☐ CURRICULUM VITAE   ☐ OTHER STATEMENT OF QUALIFICATIONS

3. NAME AND ADDRESS OF ANY MEDICAL SCHOOL, HOSPITAL OR OTHER RESEARCH FACILITY WHERE THE CLINICAL INVESTIGATION(S) WILL BE CONDUCTED.

4. NAME AND ADDRESS OF ANY CLINICAL LABORATORY FACILITIES TO BE USED IN THE STUDY.

5. NAME AND ADDRESS OF THE INSTITUTIONAL REVIEW BOARD (IRB) THAT IS RESPONSIBLE FOR REVIEW AND APPROVAL OF THE STUDY(IES).

6. NAMES OF THE SUBINVESTIGATORS *(e.g., research fellows, residents, associates)* WHO WILL BE ASSISTING THE INVESTIGATOR IN THE CONDUCT OF THE INVESTIGATION(S).

7. NAME AND CODE NUMBER, IF ANY, OF THE PROTOCOL(S) IN THE IND FOR THE STUDY(IES) TO BE CONDUCTED BY THE INVESTIGATOR.

FORM FDA 1572 (10/99)   PREVIOUS EDITION IS OBSOLETE.   PAGE 1 OF 2

8. ATTACH THE FOLLOWING CLINICAL PROTOCOL INFORMATION:

☐ FOR PHASE 1 INVESTIGATIONS, A GENERAL OUTLINE OF THE PLANNED INVESTIGATION INCLUDING THE ESTIMATED DURATION OF THE STUDY AND THE MAXIMUM NUMBER OF SUBJECTS THAT WILL BE INVOLVED.

☐ FOR PHASE 2 OR 3 INVESTIGATIONS, AN OUTLINE OF THE STUDY PROTOCOL INCLUDING AN APPROXIMATION OF THE NUMBER OF SUBJECTS TO BE TREATED WITH THE DRUG AND THE NUMBER TO BE EMPLOYED AS CONTROLS, IF ANY; THE CLINICAL USES TO BE INVESTIGATED; CHARACTERISTICS OF SUBJECTS BY AGE, SEX, AND CONDITION; THE KIND OF CLINICAL OBSERVATIONS AND LABORATORY TESTS TO BE CONDUCTED; THE ESTIMATED DURATION OF THE STUDY; AND COPIES OR A DESCRIPTION OF CASE REPORT FORMS TO BE USED.

9. COMMITMENTS:

I agree to conduct the study(ies) in accordance with the relevant, current protocol(s) and will only make changes in a protocol after notifying the sponsor, except when necessary to protect the safety, rights, or welfare of subjects.

I agree to personally conduct or supervise the described investigation(s).

I agree to inform any patients, or any persons used as controls, that the drugs are being used for investigational purposes and I will ensure that the requirements relating to obtaining informed consent in 21 CFR Part 50 and institutional review board (IRB) review and approval in 21 CFR Part 56 are met.

I agree to report to the sponsor adverse experiences that occur in the course of the investigation(s) in accordance with 21 CFR 312.64.

I have read and understand the information in the investigator's brochure, including the potential risks and side effects of the drug.

I agree to ensure that all associates, colleagues, and employees assisting in the conduct of the study(ies) are informed about their obligations in meeting the above commitments.

I agree to maintain adequate and accurate records in accordance with 21 CFR 312.62 and to make those records available for inspection in accordance with 21 CFR 312.68.

I will ensure that an IRB that complies with the requirements of 21 CFR Part 56 will be responsible for the initial and continuing review and approval of the clinical investigation. I also agree to promptly report to the IRB all changes in the research activity and all unanticipated problems involving risks to human subjects or others. Additionally, I will not make any changes in the research without IRB approval, except where necessary to eliminate apparent immediate hazards to human subjects.

I agree to comply with all other requirements regarding the obligations of clinical investigators and all other pertinent requirements in 21 CFR Part 312.

INSTRUCTIONS FOR COMPLETING FORM FDA 1572
STATEMENT OF INVESTIGATOR:

1. Complete all sections. Attach a separate page if additional space is needed.

2. Attach curriculum vitae or other statement of qualifications as described in Section 2.

3. Attach protocol outline as described in Section 8.

4. Sign and date below.

5. FORWARD THE COMPLETED FORM AND ATTACHMENTS TO THE SPONSOR. The sponsor will incorporate this information along with other technical data into an Investigational New Drug Application (IND).

10. SIGNATURE OF INVESTIGATOR	11. DATE

(**WARNING:** A willfully false statement is a criminal offense. U.S.C. Title 18, Sec. 1001.)

Public reporting burden for this collection of information is estimated to average 100 hours per response, including the time for reviewing instructions, searching existing data sources, gathering and maintaining the data needed, and completing reviewing the collection of information. Send comments regarding this burden estimate or any other aspect of this collection of information, including suggestions for reducing this burden to:

Food and Drug Administration	Food and Drug Administration	"An agency may not conduct or sponsor, and a
CBER (HFM-99)	CDER (HFD-94)	person is not required to respond to, a
1401 Rockville Pike	5516 Nicholson Lane	collection of information unless it displays a
Rockville, MD 20852-1448	Kensington, MD 20895	currently valid OMB control number."

Please **DO NOT RETURN** this application to this address.

FORM FDA 1572 (10/99)

Example of Form 1, FDA 1572, Completed

DEPARTMENT OF HEALTH AND HUMAN SERVICES
PUBLIC HEALTH SERVICE
FOOD AND DRUG ADMINISTRATION

STATEMENT OF INVESTIGATOR
(TITLE 21, CODE OF FEDERAL REGULATIONS (CFR) PART 312)
(See instructions on reverse side.)

Form Approved: OMB No. 0910-0014.
Expiration Date: September 30, 2002.
See OMB Statement on Reverse.

NOTE: No investigator may participate in an investigation until he/she provides the sponsor with a completed, signed Statement of Investigator, Form FDA 1572 (21 CFR 312.53(c)).

---

1. NAME AND ADDRESS OF INVESTIGATOR

   Include the facility address where the physician is affiliated.
   Include any co-PIs (rare).
   A separate 1572 needs to be completed for each PI.

2. EDUCATION, TRAINING, AND EXPERIENCE THAT QUALIFIES THE INVESTIGATOR AS AN EXPERT IN THE CLINICAL INVESTIGATION OF THE DRUG FOR THE USE UNDER INVESTIGATION. ONE OF THE FOLLOWING IS ATTACHED.

   [X] CURRICULUM VITAE   [ ] OTHER STATEMENT OF QUALIFICATIONS

3. NAME AND ADDRESS OF ANY MEDICAL SCHOOL, HOSPITAL OR OTHER RESEARCH FACILITY WHERE THE CLINICAL INVESTIGATION(S) WILL BE CONDUCTED.

   Important: List the names and addresses of any facility that will be evaluating or performing tests on a subject.

   EXAMPLE →

   ABC Research Centers of America
   1234 South Park Road
   Anytown, USA

   ABC Research Centers of America
   4455 West 50th Avenue
   Anytown, USA

   Rehab Associates
   4234 West 50th Avenue
   Anytown, USA

4. NAME AND ADDRESS OF ANY CLINICAL LABORATORY FACILITIES TO BE USED IN THE STUDY.

   Important: List the names and addresses of each laboratory processing laboratory samples including a local lab, if applicable, when using a central lab. Include the name and address of any facility reviewing ECGs and CT scans, if applicable.

   Central Research Laboratories
   1254 South Park Road
   Anytown, USA

   Radiology Associates
   5425 SE 5th Avenue
   Irvine, California

5. NAME AND ADDRESS OF THE INSTITUTIONAL REVIEW BOARD (IRB) THAT IS RESPONSIBLE FOR REVIEW AND APPROVAL OF THE STUDY(IES).

   Institutional Review of Anytown
   4455 South 8th Street
   Anytown, USA

   If the sponsor has contracted with a central IRB, you are not obligated to use the central IRB if you have a local IRB. Check with the local IRB to see if their schedule will accommodate the sponsor's timelines.

6. NAMES OF THE SUBINVESTIGATORS (e.g., research fellows, residents, associates) WHO WILL BE ASSISTING THE INVESTIGATOR IN THE CONDUCT OF THE INVESTIGATION(S).

   List all sub-investigators in this section. Check with the sponsor to see if the names of additional research staff need to be included, i.e., CRC, pharmacist.

   Allan Jones, MD
   John Allen, MD
   Alex Campbell, MD
   Marc Ackerman, DO
   Sally Chapps, RN, CRC
   Michael Sams, Phar. D.

7. NAME AND CODE NUMBER, IF ANY, OF THE PROTOCOL(S) IN THE IND FOR THE STUDY(IES) TO BE CONDUCTED BY THE INVESTIGATOR.

   Make sure you list the complete title of the protocol, include the IND# if available.

   Note: If the sponsor provides the Form FDA 1572 on a computer disk, make sure you make the form print out as ONE page, not two.

FORM FDA 1572 (10/99)   PREVIOUS EDITION IS OBSOLETE.   PAGE 1 OF 2

8. ATTACH THE FOLLOWING CLINICAL PROTOCOL INFORMATION:

☐ FOR PHASE 1 INVESTIGATIONS, A GENERAL OUTLINE OF THE PLANNED INVESTIGATION INCLUDING THE ESTIMATED DURATION OF THE STUDY AND THE MAXIMUM NUMBER OF SUBJECTS THAT WILL BE INVOLVED.

☐ FOR PHASE 2 OR 3 INVESTIGATIONS, AN OUTLINE OF THE STUDY PROTOCOL INCLUDING AN APPROXIMATION OF THE NUMBER OF SUBJECTS TO BE TREATED WITH THE DRUG AND THE NUMBER TO BE EMPLOYED AS CONTROLS, IF ANY; THE CLINICAL USES TO BE INVESTIGATED; CHARACTERISTICS OF SUBJECTS BY AGE, SEX, AND CONDITION; THE KIND OF CLINICAL OBSERVATIONS AND LABORATORY TESTS TO BE CONDUCTED; THE ESTIMATED DURATION OF THE STUDY; AND COPIES OR A DESCRIPTION OF CASE REPORT FORMS TO BE USED.

9. COMMITMENTS:

I agree to conduct the study(ies) in accordance with the relevant, current protocol(s) and will only make changes in a protocol after notifying the sponsor, except when necessary to protect the safety, rights, or welfare of subjects.

I agree to personally conduct or supervise the described investigation(s).

I agree to inform any patients, or any persons used as controls, that the drugs are being used for investigational purposes and I will ensure that the requirements relating to obtaining informed consent in 21 CFR Part 50 and institutional review board (IRB) review and approval in 21 CFR Part 56 are met.

I agree to report to the sponsor adverse experiences that occur in the course of the investigation(s) in accordance with 21 CFR 312.64.

I have read and understand the information in the investigator's brochure, including the potential risks and side effects of the drug.

I agree to ensure that all associates, colleagues, and employees assisting in the conduct of the study(ies) are informed about their obligations in meeting the above commitments.

I agree to maintain adequate and accurate records in accordance with 21 CFR 312.62 and to make those records available for inspection in accordance with 21 CFR 312.68.

I will ensure that an IRB that complies with the requirements of 21 CFR Part 56 will be responsible for the initial and continuing review and approval of the clinical investigation. I also agree to promptly report to the IRB all changes in the research activity and all unanticipated problems involving risks to human subjects or others. Additionally, I will not make any changes in the research without IRB approval, except where necessary to eliminate apparent immediate hazards to human subjects.

I agree to comply with all other requirements regarding the obligations of clinical investigators and all other pertinent requirements in 21 CFR Part 312.

INSTRUCTIONS FOR COMPLETING FORM FDA 1572
STATEMENT OF INVESTIGATOR:

1. Complete all sections. Attach a separate page if additional space is needed.

2. Attach curriculum vitae or other statement of qualifications as described in Section 2.

3. Attach protocol outline as described in Section 8.

4. Sign and date below.

5. FORWARD THE COMPLETED FORM AND ATTACHMENTS TO THE SPONSOR. The sponsor will incorporate this information along with other technical data into an Investigational New Drug Application (IND).

10. SIGNATURE OF INVESTIGATOR	11. DATE

(WARNING: A willfully false statement is a criminal offense. U.S.C. Title 18, Sec. 1001.)

Public reporting burden for this collection of information is estimated to average 100 hours per response, including the time for reviewing instructions, searching existing data sources, gathering and maintaining the data needed, and completing reviewing the collection of information. Send comments regarding this burden estimate or any other aspect of this collection of information, including suggestions for reducing this burden to:

Food and Drug Administration
CBER (HFM-99)
1401 Rockville Pike
Rockville, MD 20852-1448

Food and Drug Administration
CDER (HFD-94)
5516 Nicholson Lane
Kensington, MD 20895

"An agency may not conduct or sponsor, and a person is not required to respond to, a collection of information unless it displays a currently valid OMB control number."

Please **DO NOT RETURN** this application to this address.

FORM FDA 1572 (10/99)

Form 2. THE ADMINISTRATIVE CHECKLIST

Protocol # _____

Protocol Title _____

_____

Sponsor _____

Test Article _____

Clinical Research Associate _____

Phone # _____ Fax # _____

E-mail Address _____

Name of Laboratory _____

Phone # _____ Fax # _____

Institutional Review Board (IRB) _____

Chairperson _____

Phone # _____ Fax # _____

E-mail Address _____

Protocol Approval Date _____ Informed Consent Date _____

Principal Investigator _____

Sub-Investigator(s) _____

_____

Study Coordinator _____

Study Initiation Date _____ Investigator Meeting Date _____

Documents on File

- ___ Form FDA 1572
- ___ CVs/Medical License
- ___ Laboratory Certification
- ___ Complete Signed Protocol
- ___ Protocol Signature Page
- ___ Protocol Amendments
- ___ IRB Membership List
- ___ IRB Approval Letter
- ___ IRB Approved Informed Consents
- ___ Original Signed Consents
- ___ *Investigator Brochure*
- ___ Drug Accountability
- ___ Enrollment Logs
- ___ Investigator Meeting Agenda
- ___ Confidentiality Agreement
- ___ Miscellaneous Sponsor Forms

- ___ Completed Financial Disclosure Form
- ___ Serious Adverse Event Form
- ___ Laboratory Normals/Values
- ___ MedWatch Forms
- ___ Site Signature Log
- ___ Laboratory Submission Reports
- ___ Monitor Site Visit Log
- ___ Informed Consent Approval Letter
- ___ Telephone Log
- ___ Letter of Indemnification
- ___ Shipping Invoices for ALL Supplies
- ___ On-going Correspondence
- ___ Informed Consent Logs
- ___ Investigator/Meeting Attendee List
- ___ Delegation of Responsibility
- ___ Other Source Documents

NOTE:
All financial records including the following **SHOULD NOT** be part of the Site Binder, and should be kept separate:

- ✔ study budget;
- ✔ signed Study Agreement with financial compensation;
- ✔ signed Financial Disclosure Form.

However, these can be requested by the FDA during an audit.

Form 3. INITIAL ADVERSE EXPERIENCE REPORT SHEET

Date _____

Study Protocol # _____

Study Investigator _____

Subject I.D. (initials and #) _____

Date subject entered study _____

If subject discontinued study:

Date drug discontinued _____

Date subject discontinued study _____

Description of adverse experience _____

_____

*Diagnostic procedure(s) performed _____

_____

Relationship to study drug (not related, possibly, probably, definitely) _____

Subject's present condition _____

Was subject hospitalized? _____

Was sponsor notified? _____

Name of sponsor's representative contacted/date contacted _____

*May not be applicable

_____
Clinical Research Coordinator (*signature*)

Form 4. THE MEDWATCH ADVERSE EXPERIENCE FORM

Form 5. STUDY SITE TELEPHONE LOG

Date _____ Time _____

Conversation With _____

Affiliation _____

Telephone # _____

Regarding (Product, Study #, Subject #) _____

       ❑ I Placed Call   ❑ I Returned Call

       ❑ Party Called   ❑ Party Returned Call

_____

_____

_____

_____

_____

_____

_____

_____

_____

Is action or follow-up necesary?   ❑ No   ❑ Yes (specify) _____

_____

_____

_____

Was action taken?   ❑ No   ❑ Yes (specify) _____

_____

_____

_____

Date filed in study file _____

Signed _____

    cc: _____

        _____

        _____

Form 6. STUDY ADVERTISEMENT OFFICE SIGN

DO YOU HAVE [*insert study indication here*]?

If you are between the ages of _____ and _____ years, and you have been diagnosed with _____ or have any of the following symptoms on a frequent basis:

- _____
- _____
- _____
- _____
- _____

- _____
- _____
- _____
- _____
- _____

[*list signs and symptoms of disease/indication being studied*]

then you may qualify for entry into a medical research program. Please contact a nurse, receptionist, or your doctor for more information regarding this free program.

Note: IRB and Sponsor approval are required before posting signs in public areas.

Form 7. SCREEN VISIT LETTER

(Note: Print on site letterhead stationery.)

Date:

RE: Protocol #

Dear _____:

This letter is to confirm our conversation regarding your potential participation in the above referenced research study being performed at our office. This study will involve people who have a diagnosis of _____. Your screening visit is scheduled for _____ at _____, with _____.

After you have reviewed the informed consent, had all your questions answered, and have agreed to participate in this clinical trial, you will be required to: (list the study requirements here). ***(Note: It is important that no study-related procedures be performed prior to the signing of the informed consent, e.g., fasting for baseline laboratory samples, wash-out of restricted medications.)***

Your participation in this clinical trial is greatly appreciated. If you have any questions regarding your potential participation in this study, do not hesitate to contact me directly at _____. I look forward to meeting with you.

Sincerely,

_____
Clinical Research Study Coordinator

Form 8. BASELINE VISIT COVER LETTER

(Note: Print on site letterhead stationery.)

Date:

RE: Protocol #

Dear _____:

This letter is to confirm the date and time of your baseline visit for participation in the _____ research study. Your appointment has been scheduled for _____ at _____, with _____.

During your baseline visit, you will be required to undergo the following study procedures:
(Note: Some examples of study-related procedures)

- ✔ complete physical exam;
- ✔ ECG;
- ✔ provide laboratory specimens; and
- ✔ chest radiograph.

In addition, you must fast for at least ___ hours prior to having your blood drawn. This means you may not have any food or drink (including water) for the given specified time period before your scheduled appointment. A sample of your urine will be obtained as well.

I look forward to working with you on this research study. If you have any questions about your participation in this research study, please do not hesitate to contact me at _____.

Your participation in this clinical trial is greatly appreciated.

Sincerely,

_____
Clinical Research Coordinator

Form 9. ON-GOING STUDY VISIT LETTER

(Note: Print on site letterhead stationery.)

Date:

RE: Protocol #

Dear _____:

This is to confirm your next clinic visit for the _____ study on _____ at _____, with _____.

During this visit, you will be required to undergo the following study-related procedures: *(Note: List study-specific procedures here.)*

Should you need to take any medication other than the study medication between your first and subsequent visit, you will be asked to keep a record of the dates, the dosages, and reason for the medication so you may report them to me during your clinic visit. In addition, should you experience any unpleasant side effects, you need to keep a record of what they are, when they started, and how long they lasted so you may also report them to me during your clinic visit.

It is important that you remember to bring with you to this visit all unused study medication, including the box and empty blister packs and study supplies, along with your patient diary (if applicable).

It is extremely important that you keep your scheduled clinic visit. If for some reason you are unable to keep this appointment, kindly contact me immediately so that we can reschedule the appointment according to the study guidelines.

Thank you for your continued participation in this research study. In the meantime, please do not hesitate to contact me at (___)_____ if you have any questions or require additional information.

Sincerely,

_____
Clinical Research Coordinator

Form 10. FINAL STUDY LETTER

(Note: Print on site letterhead stationery.)

Date:

RE: Protocol #

Dear _____:

Dr. _____ and I would like to take this opportunity to express our sincere appreciation for your participation in Protocol _____
_____.

Without participants like you, clinical research would not be possible.

Please do not hesitate to contact me in the future if you have any questions regarding your participation in this research study. Again, thank you for your participation.

Sincerely,

_____
Clinical Research Coordinator

_____
Principal Investigator

Form 11. GENERIC FINANCIAL DISCLOSURE CERTIFICATION FORM

Principal Investigator's Name _____

Protocol # _____

Investigational Product _____

IND # _____

With respect to the clinical study for the investigational product referenced above that I am conducting for (*insert sponsoring pharmaceutical company's name*), I hereby certify to the truth and accuracy of the following statements in compliance with 21 CFR part 54, with the understanding that I am certifying not only for myself as a clinical investigator, but also for my spouse and for each dependent child of mine:

I certify that:

- ✔ I have not entered into any financial arrangement with (*insert sponsoring pharmaceutical company's name*) (e.g., bonus, royalty, or other financial incentive) whereby the outcome of the clinical study could affect my compensation;

- ✔ I do not have a proprietary interest (e.g., patent, trademark, copyright, licensing agreement, etc.) in the investigational product tested in the above referenced clinical study;

- ✔ I do not have a significant equity interest (e.g., any ownership interest, stock option, or other financial interest, the value of which cannot be calculated with reference to publicly available prices) in (*insert sponsoring pharmaceutical company's name*); and

- ✔ I have not received significant payments from (*insert sponsoring pharmaceutical company's name*), on or after February 2, 1999, having a total value in excess of $25,000, other than payments for conducting the clinical study. Examples of such significant payments include, but are not limited to, grants or funding for ongoing research, compensation in the form of equipment, retainers for ongoing consultation and honoraria that are (a) paid directly to me or to the institution with which I am affiliated, and (b) in support of my activities.

or

I disclose the following *(check all boxes that apply and attach detailed information)*:

- ❑ I have entered into a financial arrangement with (*insert sponsoring pharmaceutical company's name*), whereby the value of my compensation could be influenced by the outcome of the clinical study;

❑ I am receiving significant payments from (*insert sponsoring pharmaceutical company's name*) on or after February 2, 1999, having value in excess of $25,000, other than payment for conducting the above clinical study. Examples of significant payments include, but are not limited to, grants or funding for ongoing research, compensation in the form of equipment, retainers for ongoing consultation and honoraria that are (a) paid directly to me or to the institution with which I am affiliated, and (b) paid in support of my activities;

❑ I hold a proprietary interest (e.g., patent, trademark, copyright, licensing agreement, etc.) in the investigational product being tested in the clinical study; and

❑ I have a significant equity interest (e.g., any ownership interest, stock option, or other financial interest, the value of which cannot be calculated with reference to publicly available prices) in (*insert sponsoring pharmaceutical company's name*).

This certification shall apply throughout the entire term of the clinical study and for one year following completion of the clinical study. If there is any change in the accuracy of the foregoing statements during such time period, I shall hereby agree that I will promptly notify (*insert sponsoring pharmaceutical company's name*).

_____   _____
Signature of Principal Investigator   Date

*Tax ID#/Social Security # _____

Affiliated Institution _____

Tax ID# _____

Required for internal tracking only and will not be submitted to the FDA.

Form 12. STOP: BEFORE ANY MED CHANGES

STOP!

BEFORE ANY MED CHANGES
KINDLY CONTACT:

(INSERT NAME AND NUMBER OF PERSON RESPONSIBLE)

THIS PATIENT IS A PARTICIPANT IN A RESEARCH PROJECT

Form 13. STOP: THIS PATIENT IS A PARTICIPANT IN A RESEARCH PROJECT

STOP!

THIS PATIENT IS A PARTICIPANT IN A RESEARCH PROJECT

KINDLY CONTACT:
(INSERT NAME AND NUMBER OF PERSON RESPONSIBLE)

REGARDING THE FOLLOWING:

- ✔ PRIOR TO ANY CHANGES TO CURRENT MEDICATION
- ✔ <u>ANY</u> CHANGES IN THE PATIENT'S PHYSICAL CONDITION
- ✔ DATE AND TIME PATIENT WILL BE DISCHARGED

Form 14. PRINCIPAL INVESTIGATOR DELEGATION OF RESPONSIBILITIES

PRINCIPAL INVESTIGATOR DELEGATION OF RESPONSIBILITIES

Sponsor: _____
Study Title: _____
Study #: _____

Principal Investigator: _____
Site Name: _____
Site Number: _____

List the names, signatures, initials and procedure responsibilities for those individuals participating in the conduct of this trial

Procedures Responsibilities Legend (For Reference Only)

1. Physical
2. History
3. Phlebotomy
4. Collecting Safety Data
5. Collecting Efficacy Data
6. Dispense Medication(s)
7. CRF Entry
8. Scheduling Study Visits
9. Performing Special Study Procedures

Name and Title *(please print)*	Listed on 1572?	Signature	Initials	Procedure Responsibilities *(Enter Applicable Number(s) from Above)*	Date(s)
Name: Title:	☐ No ☐ Yes				
Name: Title:	☐ No ☐ Yes				
Name: Title:	☐ No ☐ Yes				
Name: Title:	☐ No ☐ Yes				
Name: Title:	☐ No ☐ Yes				

APPENDIX III: Individual State Regulatory Requirements for Conducting a Clinical Trial Using an Investigational Drug

STATE	CONTACT	REQUIRED DOCUMENTS SIGNED BY PATIENT OR PRINCIPAL INVESTIGATOR
Alabama	Board of Medical Examiner: (800) 227-2606	No specific state requirements. Follow FDA requirements.
Alaska	Board of Pharmacy: (907) 465-2489 Licensing Division: (907) 452-1514	Follow hospital policy.
Arizona	State Board of Pharmacy: (602) 255-5125	Follow IRB and/or hospital policy.
Arkansas	State Department of Health: (501) 661-2111 State Board of Pharmacy: (501) 682-0190	Follow hospital medical board policy.
California	Food and Drug Scientist: (916) 324-3992	Experimental subjects bill of rights in the language the subject is fluent and an informed consent.
Colorado	Board of Pharmacy Examiners: (303) 894-7750 Ext. 313	No specific state requirements. Follow FDA requirements.
Connecticut	Health Dept.: (860) 509-8000 Attorney General's Office: (860) 566-7334 Dept. of Consumer Protection: (860) 566-4490 Medical Quality Assurance: (860) 509-7563	No specific state requirements. Follow FDA requirements.
Delaware	Division of Public Health: (302) 739-4798	No specific state requirements. Follow FDA requirements.
D.C.	Food and Drug Administration: (301) 443-2894	No specific state requirements. Follow FDA requirements.
Florida	Dept. of Professional and Medical License: (850) 410-8300	No specific state requirements. Follow FDA requirements. State required language in advertising.
Georgia	Dept. of Professional and Medical Licensing: (404) 656-3913	No specific state requirements. Follow FDA requirements.
Hawaii	Food and Drug Inspector: (808) 586-4725	No specific state requirements. Follow FDA requirements.
Idaho	Dept. of Health/Board of Pharmacy: (208) 334-5546 Idaho State Board of Medicine: P.O. Box 83720 Boise, Idaho 83720-0058	No specific pharmacy requirements. Need to submit a letter to the director of Idaho State Board of Medicine.
Illinois	Drugs and Medical Devices Programs: (217) 785-22439	No specific state requirements. Follow FDA requirements.

Iowa	Food and Drug Administration: (319) 523-6381	No specific state requirements. Follow FDA requirements.
Kansas	Food and Drug Administration: (913) 752-2120	No specific state requirements. Follow FDA requirements.
Louisiana	State Food and Drug Branch: (504) 568-5401	No specific state requirements. Follow FDA requirements.
Maine	Dept. of Health: (207) 287-3201	No specific state requirements. Follow FDA requirements.
Maryland	Attorney General: (410) 767-1879	No specific state requirements. Follow FDA requirements.
Massachusetts	State Department of Health: (617) 983-6736	PI registration for 2 licenses for controlled substances—1 for practice, 1 for research; annual registration fee $50; FDA 1572 required; CV and IRB approval. Only 7 external IRB's approved.
Michigan	Board of Medicine: (517) 335-0918	No specific state requirements. Follow FDA requirements.
Minnesota	Pharmacy Board: (612) 642-0541	No specific state requirements. Follow FDA requirements.
Mississippi	Board of Pharmacy: (612) 642-0541 Medical Licensure Board: (601) 354-7190	No specific state requirements. Follow FDA requirements.
Missouri	Board of Pharmacy: (573) 751-6400	Has Dept. IRB involvement. No specific state requirements. Follow FDA requirements. Bill of Rights for patients is being worked on, not yet finalized.
Montana	Board of Pharmacy—Legal Dept.: (573) 751-6400	No specific state requirements. Follow FDA requirements.
Nebraska	Board of Pharmacy: (406) 444-4290	No specific state requirements. Follow FDA requirements.
Nevada	Board of Pharmacy: (702) 486-7380	No specific state requirements. Nevada Board of Pharmacy periodically inspects the federal documents. They require any PI administering investigational drugs to have a dispensing license under the State Pharmacy Law. A PI is subject to periodic inspections. All federal documentation must be in place.
New Hampshire	Food and Drug Administration: (800) 891-8295	No specific state requirements. Follow FDA requirements.
New Jersey	Food and Drug Administration: (800) 532-4440	No specific state requirements. Follow FDA requirements.
New Mexico	U of NM School of Medicine: (505) 272-2321	No specific state requirements. Follow FDA requirements.
New York	Dept. of Health: (518) 474-5014	Voluntary Informed Consent, and each public or private institution or agency that conducts or that proposes to conduct or authorize human research, shall establish a research review board.
North Carolina	State Pharmacy Board: (919) 515-5040	No specific sate requirements. Follow FDA requirements.
North Dakota	State Pharmacy Board: (701) 328-9535	Follow FDA requirements. Company must be a licensed wholesale drug distributor/manufacturer. Pharmacy board would like to be notified before a study begins.

Ohio	State Pharmacy Board: (614) 466-3543	Follow FDA requirements. Pharmaceutical company must have license as wholesaler, and PI or pharmacist can dispense the drug.
Oklahoma	Oklahoma State Health Dept.: (405) 271-5070 Oklahoma State Health Pharmacy: (405) 271-1959 Oklahoma Medical Association (405) 843-9571	No specific state requirements. Follow FDA requirements.
Oregon	Public Affairs Board of Pharmacy: (503) 731-4032	No specific state requirements. Follow FDA requirements.
Pennsylvania	State Public Health Dept. (717) 787-2307	No specific state requirements. Follow FDA requirements.
Rhode Island	Drug Control. Dept. of Health: (401) 277-2837	No specific state requirements. Follow FDA requirements.
South Carolina	Board of Pharmacy: (803) 737-3981	No specific state requirements. Follow FDA requirements.
South Dakota	Dept. of Health: (605) 773-3361	No specific state requirements. Follow FDA requirements. If controlled substance, need to notify the Dept. of Health.
Tennessee	FDA (Nashville): (615) 781-5374	No specific state requirements. Follow FDA requirements.
Texas	State FDA, Drug Division: (512) 719-0237	No specific state requirements. Follow FDA requirements.
Utah	U of Utah Medical Center: (801) 585-2185	No specific state requirements. Follow FDA requirements. Individual hospitals have their own policies that must be met.
Vermont	Board of Pharmacy: (802) 828-2875	Specific form and fee may be required.
Virginia	Dept. of Health: (804) 786-6272	No specific state requirements. Follow FDA requirements.
Washington	Dept. of Professional and Medical License: (360) 586-5846	No specific state requirements. Follow FDA requirements.
West Virginia	Health and Human Resources: (304) 558-0684	No specific state requirements. Follow FDA requirements.
Wisconsin	Pharmacy Board/Public Health Dept: (608) 266-0722	No specific state requirements. Follow FDA requirements.
Wyoming	Public Health Dept.: (307) 777-7172 Policy Board: (307) 777-6186	No specific state requirements. Follow FDA requirements.

APPENDIX IV: Conversion Tables

A. Conversion to Military Time

Military time is simple and logical. It is based on the fact that each day is composed of 24 hours numbered from 1 through 24. As with conventional time, the last two digits of military time are used to indicate the minute after the hour. For example, the conventional times 6:30 am and 11:45 pm would be written in military times as 0630 and 2345, respectively.

1:00 am = 0100	1:00 pm = 1300
2:00 am = 0200	2:00 pm = 1400
3:00 am = 0300	3:00 pm = 1500
4:00 am = 0400	4:00 pm = 1600
5:00 am = 0500	5:00 pm = 1700
6:00 am = 0600	6:00 pm = 1800
7:00 am = 0700	7:00 pm = 1900
8:00 am = 0800	8:00 pm = 2000
9:00 am = 0900	9:00 pm = 2100
10:00 am = 1000	10:00 pm = 2200
11:00 am = 1100	11:00 pm = 2300
12:00 pm = 1200	12:00 am = 2400

IMPORTANT: Note that 12:00 midnight is written 2400 in military time. However, one never goes past 2400 hours. One minute past midnight becomes 0001, thirty minutes past midnight becomes 0030, etc., right up to fifty-nine minutes past midnight, which is 0059. The time 1:00 am then becomes 0100.

B. Measures and Equivalents

Volume

    29.57 mL = 1 fluid ounce
    473 mL = 1 pint
    1 liter = 33.8 fluid ounces
    3785 mL = 1 gallon
    5 mL = 1 teaspoon
    15 mL = 1 tablespoon or 1/2 fluid ounce

Weight

    1 kilogram [kg] = 2.2 lbs.
    1 pound = 453.59 grams
    1 ounce = 28.35 grams
    1 gram = 15.432 grains
    1 grain = 65 mg
    1/2 grain = 32.4 mg
    1/4 grain = 16.2 mg

Height

    1 centimeter [cm] = 0.39 inches
    1 inch = 2.54 cm
    1 meter = 39.37 inches
    1 micron = 1/1000 mm

Other useful measures

    1 liter of water weighs 1 kg
    1 mL of water weighs 1 gram

C. Conversion Charts for Height and Weight Measurements

Feet Inches	Inches	Centimeters
4' 8"	56	142.2
4' 9"	57	144.8
4' 10"	58	147.3
4' 11"	59	149.9
5' 0"	60	152.4
5' 1"	61	154.9
5' 2"	62	157.5
5' 3"	63	160.0
5' 4"	64	162.6
5' 5"	65	165.1
5' 6"	66	167.6
5' 7"	67	170.2
5' 8"	68	172.7
5' 9"	69	175.3
5' 10"	70	177.8
5' 11"	71	180.3
6' 0"	72	182.9
6' 1"	73	185.4
6' 2"	74	188.0
6' 3"	75	190.5
6' 4"	76	193.0
6' 5"	77	195.6
6' 6"	78	198.1
6' 7"	79	200.7
6' 8"	80	203.2

Pounds	Kilograms
1	0.4535
2.2046	1.0
95	43.1
100	45.4
105	47.6
110	49.9
115	52.2
120	54.4
125	56.7
130	59.0
135	61.2
140	63.5
145	65.8
150	68.0
155	70.3
160	72.6
165	74.8
170	77.1
175	79.4
180	81.6
185	83.9
190	86.2
195	88.5
200	90.7
205	93.0
210	95.3
215	97.5
220	99.8
225	102.1
230	104.3
235	106.6
240	108.9
245	111.1
250	113.4
255	115.7
260	117.9
265	120.2
270	122.5
275	124.7
280	127.0
285	129.3
290	131.5
295	133.8
300	136.1

APPENDIX V: Glossary of Terms

Adverse Experience	A medical complaint, change, or possible side effect of any degree of severity, that may or may not be attributed to the test article.
CRA	*Clinical Research Associate.* The Sponsor's representative who is responsible for monitoring administrative aspects of the study. The CRA should be contacted when any question regarding study procedures or paperwork arises.
CRF	*Case Report Form.* Standardized set of data input forms onto which all information pertaining to the study is recorded.
CRO	*Clinical Research Organization.* An organization contracted by the Sponsor to conduct or monitor various aspects of the study.
Crossover Study	A study in which each patient receives all or some of the test drugs during different phases of the study.
Double-Blind Study	A study design in which neither the investigator nor the patient know if the drug, comparator, or placebo is being used.
IDE	*Investigational Device Exemption.* This document provides sufficient data to establish that a device has demonstrated a reasonable degree of safety that does not preclude its testing in humans.
IND	*Investigational New Drug Application.* Also known as the Notice of Claimed Investigational Exemption for a new drug, this document provides sufficient data to establish that the drug has demonstrated a reasonable degree of safety that does not preclude its testing in humans.
Investigator Brochure	A document that summarizes everything known about the test article to date; also known as the Clinical Brochure. One of its purposes is to inform investigators what medical actions and side effects have occurred in past studies and to define the pharmacology and toxicology of the drug based on animal studies.
NDA	*New Drug Application.* A comprehensive compilation of the documentation used to support the safe and effective use of a drug.
Open-Label Study	A study in which both the investigator and patient know the identity of the test drug.

PI	***Principal Investigator.*** The individual who is ultimately responsible for the conduct of the clinical investigation; the person under whose immediate direction the test article is administered or dispensed.
Protocol	The document that gives specific instructions for the conduct of the study.
Single-Blind Study	A study in which the investigator, but not the patient, knows the identity of the test drug.
Source Documents	Any record on which information was originally recorded. Examples of source documents are case report forms, laboratory reports, hospital records, and notes kept in the study file. When entries on CRFs are taken from other documents such sources must be available for review by the CRA and other personnel authorized by the Sponsor.
Sponsor	The company or organization that is paying for the conduct of a clinical study; usually the pharmaceutical company or device manufacturer whose drug or device is being tested.
Test Article	A drug, placebo, food additive, or medical device being investigated for human use.

Types of Clinical Studies

Phase I:	Initial studies to evaluate the safety of the drug in a single dose in a small number of healthy volunteers.
Phase II:	Designed to evaluate the preliminary efficacy of the drug in a small number of patients with the target disease. Also used to better define the dose.
Phase III:	Large-scale double-blind studies in large numbers of patients designed to accurately determine the efficacy and safety profile of the drug.
Phase IV:	Post-marketing surveillance studies and other special trials designed to further detail the safety and efficacy of the drug.
Special Studies:	Done in special populations [the elderly, those with renal or hepatic impairment]; drug interaction studies and other unique studies.

Other Books of Interest

Genetics & Your Health: A Guide for the 21st Century Family
By Raye Lynn Alford, Ph.D., FACMG

Public interest in genetics has never been greater now that gene research promises to revolutionize medicine in the 21st century. In addition to the medical applications, the confidentiality of information and regulation of genetic technologies are hot-button topics. *Genetics and Your Health* will answer your questions about what the startling advances in genetic research, testing, and therapy really mean to today's family. Included is a directory to medical resources for genetics care, support, and information over the Internet, and the latest word on the Human Genome Project.

Medford Press/Plexus
1999/266 pp/softbound
ISBN 0-9666748-1-2
$19.95

Medford Press/Plexus
1999/266 pp/hardbound
ISBN 0-9666748-2-0
$29.95

Glossary of Terms Used in Informed Consent Documents for Pharmaceutical Trials: A Concise Reference for Clinical Researchers
By Deborrah Norris

This invaluable reference provides more than 1500 medical terms and their lay language definitions. The book also includes a detailed, thoroughly researched history and analysis of the informed consent document, as well as explicit instructions on how to compose such a document.

"The important contribution of this book is that it will facilitate translation of technical terms to more generally accessible language. In this way, it will assist clinicians, investigators, members of institutional review boards, and others in their efforts to improve the process of informed consent."

—Robert J. Levine, MD
Author, *Ethics and Regulation of Clinical Research*

Plexus Publishing, Inc. • 1996/70 pp/softbound • 0-9631310-3-6 • $39.95

Super Searchers on Health & Medicine:
The Online Secrets of Top Health & Medical Researchers
By Susan M. Detwiler; Reva Basch, editor

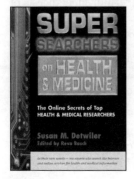

Susan M. Detwiler brings together ten of the world's leading researchers for the benefit of healthcare professionals who need to locate, evaluate, and use online health and medical information. Through the author's probing interviews, medical librarians, clinical researchers, health information specialists, and physicians reveal how they combine traditional sources with the best of the Internet. If you use online databases for critical health and medical research, these Super Searchers will help guide you around the perils and pitfalls to the most important services, sources, and techniques. As a reader bonus, recommended Internet resources are only a click away at "The Super Searchers Web Page."

Super Searchers on Health & Medicine—edited by noted author and online searcher Reva Basch—is the fourth title in the new Super Searchers series for today's serious information user.

Information Today, Inc. • 2000/200 pp/softbound • ISBN 0-910965-44-7 • $24.95

Other Books of Interest

Pharmacology of Therapeutic Agents Used in Anesthesia: An Introduction
By Robert I. Katz, Stephen A. Vitkun, and Thomas R. Eide

Here is a practical guide to all classes of therapeutic agents the anesthesiologist may encounter or administer during the perioperative period. This concise reference elucidates the pharmacology of each class of drugs and provides the anesthesiologist with the basic footing needed to understand the clinical implications of administering non-anesthetic agents in conjunction with anesthesia. This substantive yet accessible text is one from which both practicing anesthesiologists and residents can learn—and relearn—important principles of pharmacology in patient care.

"An eminently readable guide to the applied pharmacology of anesthesia. I highly recommend it."

—Albert J. Saubermann, MD
Professor and Chairman/Department of Anesthesiology
Albert Einstein College of Medicine of Yeshiva University
Montefiore Medical Center

Plexus Publishing, Inc. • 1994/311 pp/hardbound • 0-9631310-2-8 • $68.00

A Dictionary of Natural Products
By George Macdonald Hocking, Ph.D.

A Dictionary of Natural Products is primarily devoted to an arrangement and explanation of terms relating to natural, non-artificial crude drugs from the vegetable, animal, and mineral kingdoms. This volume presents over 18,000 entries of medicinal, pharmaceutical, and related products appearing on the market as raw materials or occurring in drug stores, folk medical practice, and in chemical manufacturing processes.

Plexus Publishing, Inc. • 1997/1,024 pp/hardbound • ISBN 0-937548-31-6 • $139.50
(A $6 shipping and handling fee will be added to the cost.)

Biology Digest

Biology Digest is a comprehensive abstracts journal covering all the life sciences. Each monthly issue contains over 300 abstracts which are, in essence, individual digests of articles and research reports gathered from worldwide sources. Important information is retained in the abstracts to give a precise, inclusive summary of the original material.

Biology Digest was specially created to meet the needs of high school and undergraduate college students. It provides easy access to new scientific developments at a comprehension level appropriate for students. However, *Biology Digest* has proved to be useful to biologists at all levels—professional and amateur alike.

Biology Digest is also available on CD-ROM, which is updated monthly and covers 1989 to the present. For more information on the CD-ROM version, contact NewsBank, Inc., 58 Pine Street, New Canaan, CT 06840-5426, or call 1-800-762-8182.

Plexus Publishing, Inc. • Volume 27 (2000/01) • Monthly (September-May)/ISSN 0095-2958 • 1 year $139.00 *(New subscribers qualify for the special introductory price of $109.00 or may purchase volumes 26 & 27 as the "Get Started Package" for $186.00.)*

Notes

**For a publications catalog,
call 609-654-6500, or
log onto www.plexuspublishing.com**

Notes